Astronomy Explained

520 NOR
(3 wks)

Springer

London
Berlin
Heidelberg
New York
Barcelona
Budapest
Hong Kong
Milan
Paris
Santa Clara
Singapore
Tokyo

Astronomy Explained

Gerald North

With 161 Figures

 Springer

Gerald North BSc
9 Camperdown Street, Sidley, Bexhill-on-Sea,
East Sussex TN39 5BE, UK

Cover illustrations: Front cover image of Jupiter was taken with the Hubble Space Telescope's Planetary Camera and shows eight impact sites from the disintegrated comet Shoemaker-Levy 9 in July 1994 (courtesy Hubble Space Telescope Comet Team and NASA). Back cover photograph of Villafranca (courtesy ESA).

ISBN 3-540-76136-5 Springer-Verlag Berlin Heidelberg New York

British Library Cataloguing in Publication Data
North, Gerald
    Astronomy explained
    1.Astronomy 2.Space sciences 3.Astronomy – Problems,
    exercises, etc. 4.Space sciences – Problems, exercises, etc.
    I.Title
    520
ISBN 3540761365

Library of Congress Cataloging-in-Publication Data
North, Gerald, 1958–
    Astronomy explained / Gerald North.
       p.   cm.
    Includes index.
    ISBN 3-540-76136-5 (Berlin : pbk. : acid-free paper)
       1. Astronomy.   I. Title.
    QB45.N69   1997   97–3761
    520—DC21   CIP

© Springer-Verlag London Limited 1997
Printed in Malta
2nd printing 1998

Typeset by EXPO Holdings, Malaysia
Printed by Interprint Ltd., Malta
58/3830-54321   Printed on acid-free paper

Preface

Every year large numbers of people take up the study of astronomy, mostly at the amateur level. There are plenty of elementary books on the market, full of colourful photographs but lacking in proper explanations of how and why things are as they are. Many people eventually wish to go beyond the coffee-table book stage and study this fascinating subject in greater depth. This book is written for them.

In addition, many people sit for public examinations in this subject each year. This book is also intended to be of use to them. I have been careful to cover all the topics in the syllabus of the GCSE astronomy course and to provide a selection of sample questions at the end of each chapter. Although these questions are intended for prospective examination candidates, the more casual reader might wish to try some of them just for fun! Astronomy is also a component part of many college and schools subjects – ranging from general studies options to whole sections of physics and geology courses.

Much material is available at both the elementary and advanced (university) levels in this subject but little bridging the gap. Nevertheless, today's amateur astronomers are becoming very much more sophisticated in their approach than was the case just a few years ago. In this book I have tried to give a comprehensive treatment of the subject but in more depth than is usually found in elementary works. Some mathematics is included, though it is only at a basic level. Much of Chapter 1 covers material specified in the GCSE examination syllabus and can be glossed over by the casual reader. The same is true for the parts of Chapter 2 that are concerned with calculations on time. I hope that you find this book interesting. If you are sitting for an examination in, or involving, astronomy I wish you the very best of luck and I hope that this book will help you.

Gerald North
1997

Acknowledgements

I would like to thank the Director and staff of the Royal Observatories, Cambridge (formerly the Royal Greenwich Observatory, Herstmonceux), and the Jet Propulsion Laboratory, Pasadena, for permission to use the photographs included in this book.

In addition, my friends Martin Mobberley and Eric Strach have very kindly let me use some of their most excellent photographs.

I would also like extend my thanks to Mr John Winckler and his staff at Macmillan Press Ltd., for their sterling work in the publishing of the first edition of this book and to Mr John Watson and his staff at Springer-Verlag for their excellent work in the publication of this new and enlarged edition.

I am most grateful to all of the above.

G.N.

Contents

3   The Telescope

4   Modern Developments in Instrumentation

5   Gravitation

Figures

Figure 1.1 The constellation of Orion, photographed by the author. The brightest stars in this time-exposure photograph form the very distinctive shape that the constellation has to the naked eye. The faintest stars shown here are ordinarily much too dim to be seen by the unaided eye. The amorphous blob below the three bright stars that make up the "belt of Orion" is the gaseous nebula M42.

Chapter 1

The Celestial Sphere

THE Greek philosopher Ptolemy summarised the ancient view of the structure of the Universe in about 140 AD. This view placed the Earth at the centre of the Universe with the Sun, Moon, planets and stars attached to crystal spheres, concentrically arranged, that revolved about the Earth.

It was not until 1543 that Nicolaus Copernicus published his theory that it was the Sun that was the centre of the Universe (the so-called "heliocentric theory"), and it was not until 1610 that Galileo could provide any evidence to confirm this view. We now know that even the Sun is not the centre of the Universe – it is but one very insignificant member of a vast system of stars known as the Galaxy. Further, the Galaxy is but one of countless millions of other galaxies in the infinity of space.

1.1 The Concept of the Celestial Sphere

Owing to the enormous distances separating us from even the nearest stars, they seem to remain in constantly fixed patterns that do not alter with the seasons nor with the progress of day and night. To an observer on the Earth it appears as if the stars are fixed to some enormous dome encircling the entire Earth which is slowly rotating about its axis (completing one rotation in approximately one day). It is now known that it is the Earth and not this "star-dome" that rotates, but in positional astronomy the concept of the "star-dome" or, more properly, the *celestial sphere* is still used today.

Figure 1.2 (*overleaf*) illustrates the general principle of the celestial sphere. The starry firmament appears to rotate about two fixed points in the sky (on opposite hemispheres of the celestial sphere). These points are the projections of the Earth's poles of rotation and hence are known as the celestial poles. The north celestial pole is the projection of the Earth's North Pole and, similarly, the south celestial pole is the projection of the Earth's South Pole. The projection of the Earth's equator on the celestial sphere is known as the celestial equator.

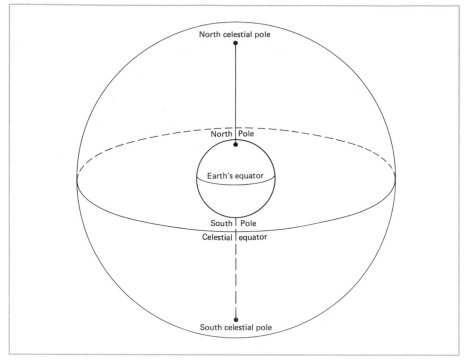

Figure 1.2 The celestial sphere.

1.2 Latitude and Longitude

In order to describe the position of any location on the surface of the Earth, a coordinate system has been set up that utilises two coordinates known as *latitude* and *longitude*. The latitude of a location is defined as its angular distance from the Earth's equator as measured from the centre of the Earth. Figure 1.3 illustrates the measurement of latitude. Latitude is defined to be **positive** if the location is **north** of the celestial equator and **negative** if the location is **south** of the equator. The latitude of the equator is thus 0° and that of the North Pole is +90°, while that of the South Pole is –90°. The latitude of London, England, is about +51° 30′.

In order to understand the measurement of longitude we must first understand the concept of the *great circle*. A great circle is a line that can be drawn around a sphere such that a circle is described whose plane passes through the centre of the sphere. It should be apparent that any great circle will cut the sphere exactly in half.

An infinite number of great circles can be drawn on any sphere. However, on the Earth one kind of great circle has special significance. Any great circle that passes through both of the Earth's poles is known as a *meridian*. A meridian cuts the equator at right angles. Because of the historical status of Greenwich as the

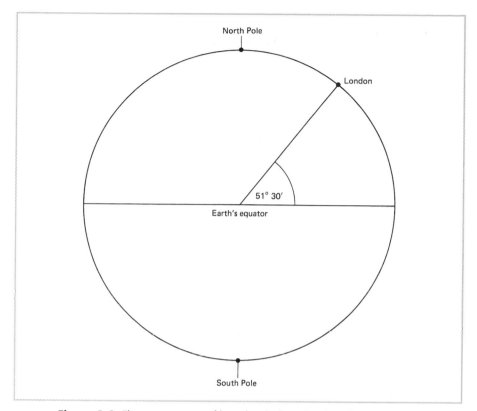

Figure 1.3 The measurement of latitude. The latitude of London is 51° 30′.

timekeeping centre of the world, due primarily to its former naval importance, it was chosen as the location from which longitude was to be measured. The great circle which passes through the North Pole, the South Pole and the Royal Observatory, Greenwich, is known as the *Greenwich meridian*.

If a location on the Earth's surface is considered to lie on its own meridian then the angular distance, measured eastwards from the centre of the Earth, along the equator, between the Greenwich meridian and the meridian of the location gives the longitude of the location. The measurement of longitude is illustrated in Figure 1.4 (*overleaf*).

1.3 Declination and Right Ascension

Just as it may be necessary to give the exact position of some object on the surface of the Earth, we may equally want to pinpoint the position of a celestial body in the sky. To do this the concept of the celestial sphere can be used (Fig. 1.5, *overleaf*). The angular distance that a body subtends from the celestial equator we call the *declination* of that body. Declination is measured in degrees, minutes

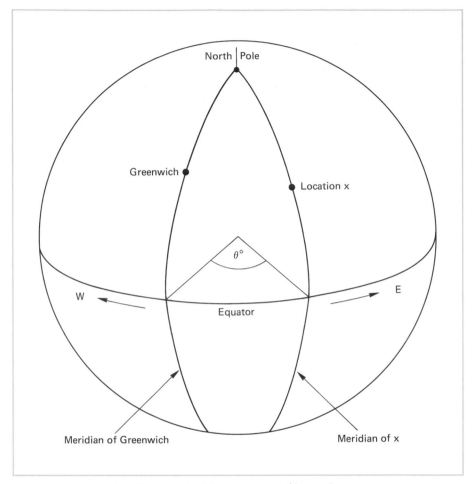

Figure 1.4 The measurement of longitude.

and seconds and is **positive** for objects **north** of the celestial equator, whilst objects **south** of the celestial equator have **negative** declinations. Thus the measurement of declination on the celestial sphere is akin to the measurement of latitude on the surface of the Earth.

To measure longitude we need a zero point from which to measure. In the case of objects on the celestial sphere astronomers have to define a similar point. For navigation on the Earth we have chosen a special location, Greenwich, in order to define our zero point of reference – the Greenwich meridian. For the celestial coordinate system astronomers have chosen a special point in the sky, known as the *vernal equinox*. This point is actually on the celestial equator and all measurements are made from it.

The angular distance between the meridian on which the celestial body lies and the vernal equinox, as measured from the Earth in an eastwards direction along

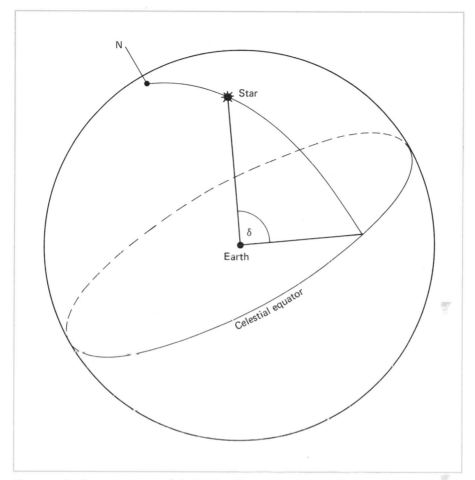

Figure 1.5 The measurement of declination. The angle δ is the declination of the star. In this example the star is north of the celestial equator and so the star has a positive declination. If the star were situated in the southern hemisphere of the sky then its declination would be negative.

the celestial equator, is called the *right ascension* of the body (see Fig. 1.6, *overleaf*). Right ascension is the celestial equivalent of longitude but, unlike our Earthly longitude, right ascension (RA) is not often measured in degrees. The Earth turns on its axis once every 23 hours 56 minutes and so any particular star will appear to go once round the sky in this time, an angular distance of 360°. As will be seen in the next chapter, our everyday twenty-four-hour clock is based upon the apparent motions of the Sun, which is not quite in step with the stars. Astronomers have devised a special time-scale which keeps pace with the stars, known as *sidereal time*.

In the astronomers' system, 24 sidereal hours is the time taken for a star to appear to go once round the sky, as viewed from the surface of the Earth (i.e. to complete a full circle of 360°). Thus a star on the celestial equator will take one

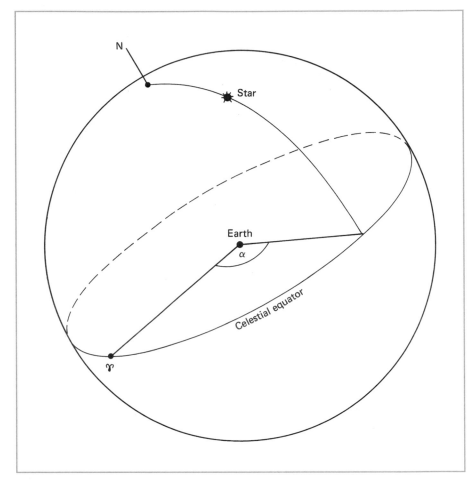

Figure 1.6 The measurement of right ascension. The angle α between the vernal equinox, ♈, and the position where the star's own meridian cuts the celestial equator (as seen from the Earth and measured eastwards along the celestial equator) is the right ascension of the star.

sidereal hour to move an angular distance of 15° across the sky (360°/24 = 15°). In astronomy it is convenient to measure right ascension in units of sidereal time instead of normal angular measure.

The symbol for right ascension is α. The symbol for declination is δ. For example, the coordinates of the star Arcturus are written as:

$$\alpha = 14^{\mathrm{h}}\ 14^{\mathrm{m}}\ 37^{\mathrm{s}}$$

$$\delta = +\ 19°\ 18'\ 05''$$

Figure 1.7 shows the position of Arcturus on the celestial sphere. It is important to remember that the vernal equinox remains fixed with respect to the stars, just as the location of Greenwich remains fixed on the Earth's surface. Both points provide fixed references for their respective coordinate systems.

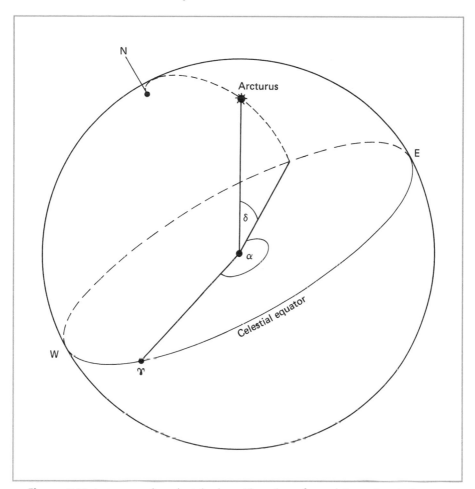

Figure 1.7 Arcturus on the celestial sphere. The values of α and δ are given in the text.

1.4 The Altitude of the Celestial Equator and Poles

With reference to Figure 1.8 (*overleaf*), it is obvious that the north celestial pole will only appear to be directly overhead to an observer standing on the Earth's North Pole. Similarly the celestial equator will appear to pass directly overhead only when the observer is actually on the Earth's equator. When an observer sees a celestial object directly overhead the object is said to be at the *zenith*. If the object is in the diametrically opposite position, directly beneath his feet and through the other side of the Earth, then the object is said to be at the *nadir*.

Great circles can be drawn on the celestial sphere that pass through both the zenith and the nadir and consequently intersect the horizon of the observer, at

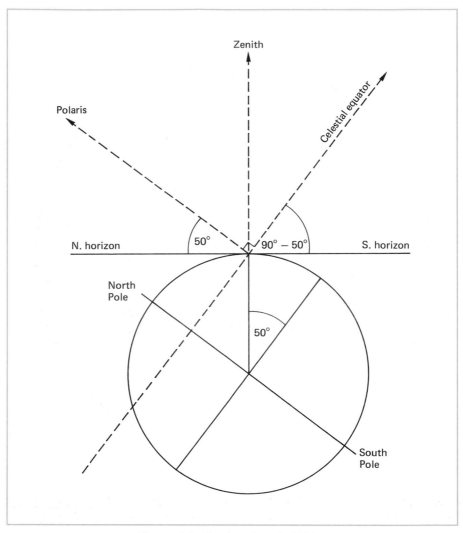

Figure 1.8 The sky at latitude 50° N.

his particular location, at right angles. If the celestial body is considered to lie on one of these great circles then the angular distance between the horizon and the body, measured from the observer along this great circle, is the *altitude* of the body (see Fig. 1.9).

Figure 1.8 shows that from, say, a latitude of +50° (50° N), the north celestial pole will not be at the zenith but will lie in the sky above the northern horizon at an altitude of 50°. The star Polaris, in the constellation of Ursa Minor, lies conveniently close to the north celestial pole and so makes a good marker. The north latitude of an observer is then (approximately) given by the altitude of Polaris.

For an observer, the direction of the north *cardinal point* is found by following the meridian on which he is situated in a northerly direction. The direction of

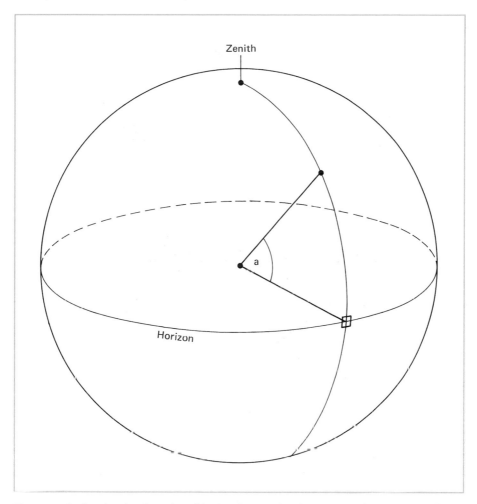

Figure 1.9 The altitude of a celestial body. The angular elevation, *a*, of a celestial body, as seen from a particular site on the surface of the Earth, is known as the altitude of the body.

the southern cardinal point is found by following the same meridian in the opposite direction. In the same way a line crossing the observer's position at right angles to his meridian gives the directions of the east and west cardinal (or geographical compass) points.

An observer in the northern hemisphere of the Earth will see that the celestial equator is inclined at an angle to the horizon, passing from the east cardinal point, reaching its greatest altitude due south, and then arching downwards to intersect the west cardinal point. The view from the southern hemisphere is reversed, with the south celestial pole above the southern horizon and the celestial equator reaching its greatest altitude due north. Incidentally, the people living in the southern hemisphere are unfortunate in having no distinctive star to mark the location of the south celestial pole. The inclination of the celestial equator to the observer's horizon is illustrated in Figure 1.10 (*overleaf*).

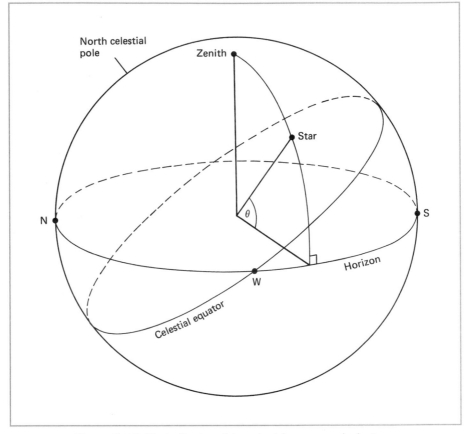

Figure 1.10 The inclination of the celestial equator to the horizon.

1.5 The Starry Vault

It is because of the Earth's west-to-east rotation that the stars appear to move together in an east-to-west direction. Thus all celestial bodies appear to rise in the east and set in the west. Since the stars are at such large distances compared with the radius of the Earth, they appear to remain essentially fixed in their positions no matter from where on Earth they are observed.

A great circle can be drawn that passes through the observer's zenith, the southern cardinal point, the nadir and the northern cardinal point. Such a great circle will obviously pass through both celestial poles and will intersect the horizon (at the cardinal points) at right angles (see Fig. 1.11). Such a great circle has a special name. It is termed the *observer's meridian*.

When a star, or other celestial body, crosses the observer's meridian it is then said to *transit* the meridian. When a star transits the meridian to the south of the north celestial pole, it will be at its greatest altitude above the horizon and it is then said to be at *upper culmination*. When the star transits the meridian to the north of the celestial pole it is then at its lowest altitude (indeed, it may well be

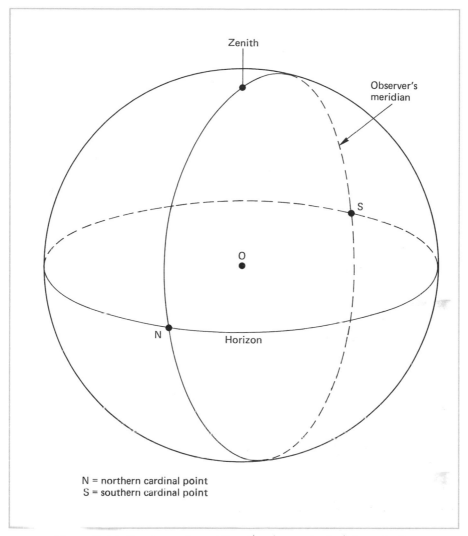

Figure 1.11 The observer's meridian. The observer is situated at point O.

below the observer's horizon) and it is then said to be at *lower culmination*. This is the case for an observer in the northern hemisphere of the Earth. In the southern hemisphere the situation is reversed.

1.6 A Better Definition of Right Ascension

The sidereal day can now be defined as the time interval between successive upper culminations of a star. Any star, or even our special point of interest on the celestial sphere – the vernal equinox – will achieve one upper culmination every sidereal day.

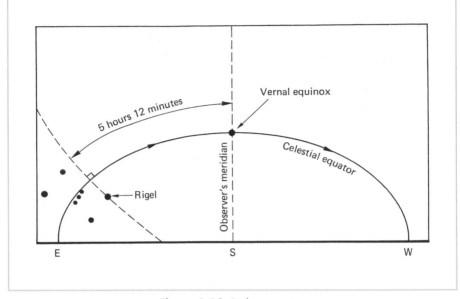

Figure 1.12 Right ascension.

Remembering that right ascension is measured using the vernal equinox as the zero point, the right ascension of a celestial body is given by the time interval between upper culminations of the vernal equinox and the object considered. Figure 1.12 illustrates the point. In the diagram the star, Rigel, has an RA of 5^h 12^m.

1.7 To Calculate the Maximum Altitude of a Star

Since any star is at upper culmination when at its greatest altitude, its altitude is then equal to its declination plus the altitude of the celestial equator where it crosses the observer's meridian (to the south of the north celestial pole) for an observer in the northern hemisphere. The altitude of the point where the celestial equator crosses the meridian is given by the *colatitude* of the observer's position. This is 90° – latitude (see Fig. 1.13).

Example 1

What is the maximum altitude of the star Betelgeux (δ = +7° 24′) from London (latitude = +51° 30′)?

Maximum altitude = (90° – 51° 30′) + (7° 24′)
                                    = 45° 54′.

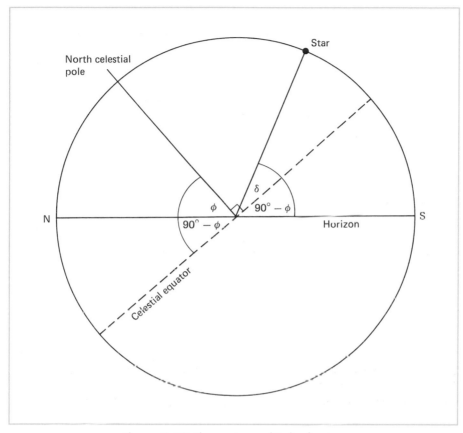

Figure 1.13 The maximum altitude of a star.

Example 2

What is the maximum altitude attained by the star Sirius (δ = –16° 41′ as seen from Paris (latitude = +48° 50′)?

You will notice that Sirius has a negative declination, or, in other words, Sirius lies to the south of the celestial equator. This fact makes no difference to the calculation.

Maximum altitude = (90° – 48° 50′) + (–16° 41′)
$$= 24° \, 29′.$$

In order to calculate the maximum altitude of a star as seen from the southern hemisphere the same equation is used but with the signs reversed. In other words, the latitude is **added** to 90° (but remember that the latitude is now a negative number) and the star's declination is now **subtracted** from the result. One more example should help to make this clear.

Example 3

What is the maximum possible altitude of the star Sirius as seen from Parkes, in New South Wales (latitude = −33° 00′)?

Maximum altitude = (90° + [−33° 00′]) − (−16° 41′)
$$= 57° 00′ + 16° 41′$$
$$= \underline{73° 41′}.$$

1.8 Circumpolar Stars

The entire starry heavens appear to rotate about the two celestial poles. It follows that stars within a certain angular distance of the visible pole (the north celestial pole from the Earth's northern hemisphere) will never set below the level horizon. It can be seen from Figure 1.14 that stars with greater declinations than the colatitude never dip below the horizon. Such stars are called *circumpolar stars*. Thus from London stars that have declinations of greater than + 38° 30′ are circumpolar.

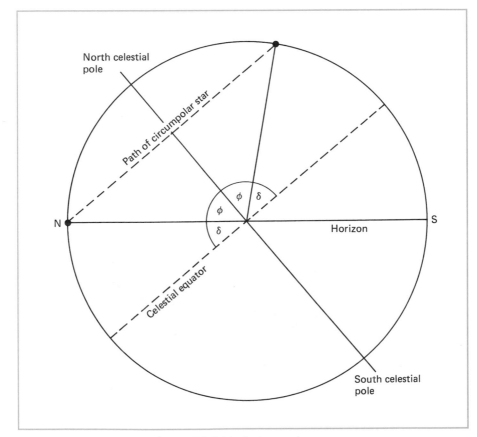

Figure 1.14 North circumpolar stars.

1.9 Stars that Never Rise

Figure 1.15 shows that, in the case of the northern hemisphere, stars with southerly (negative) declinations greater than the colatitude of the observation site can never rise above the horizon. For instance, stars further south than those with declinations – 38° 30′ can never be observed from London. In order to see these stars one would have to travel further south. Can you see what the limiting declination is for a star not to be seen from a particular location on the southern hemisphere?

1.10 Azimuth

Two coordinates, right ascension and declination, are necessary to define the position of an object on the celestial sphere. Sometimes it is more convenient to have a coordinate system that is related to the observer's local horizon.

Two coordinates are still needed. One of these is the altitude of the body above the level horizon. The other coordinate is the *azimuth*. Azimuth is measured from

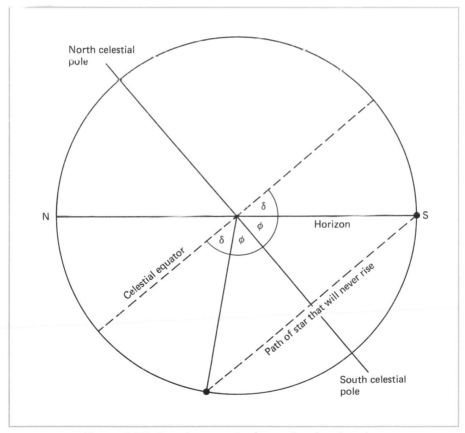

Figure 1.15 Stars that never rise from mid-northern latitudes.

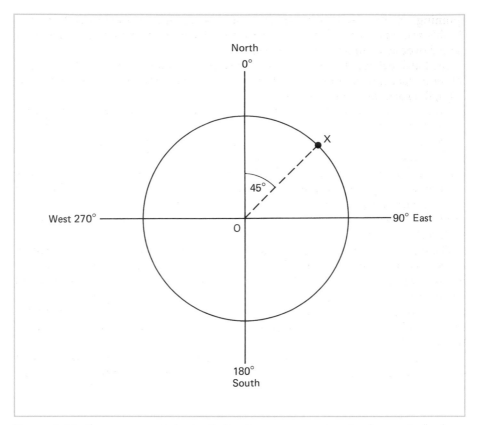

Figure 1.16 The measurement of azimuth. The diagram represents a view from vertically above the observer, who is positioned at point O. The circle represents his horizon. If the observer sees a star vertically above the point X on his horizon then he will measure its azimuth as 45°.

the north cardinal point, along the horizon, in an eastwards direction. So the azimuth of the south cardinal point is 180° and that of the west cardinal point is 270° and so on (see Fig. 1.16).

The system of coordinates that use right ascension and declination is called the *equatorial* system and the system that uses altitude and azimuth the *altazimuth* system. Though the coordinates of right ascension and declination keep pace with the stars in their apparent motions across the sky (their *diurnal motion*), the altitude and azimuth of a celestial body continually change with time. Nevertheless, the altazimuth system of coordinates is still useful.

1.11 The Earth in Space

The Earth moves through space in an approximately circular path, or *orbit*, that is centred on the Sun. The mean distance from the Earth to the Sun is 150 million km and it takes the Earth one year to go once round the Sun. The Earth is also spinning on its axis, taking one sidereal day for one complete rotation. The Earth's spin axis is not "upright" (perpendicular to the plane of its orbit) but is canted over at an angle of $23\frac{1}{2}°$ (see Fig. 1.17).

The Earth behaves like a gyroscope in that it keeps its spin axis in a fixed orientation in space, throughout its path round the Sun. Referring to Fig. 1.18 (*overleaf*), the Earth takes exactly one year to travel from position E_1 back to the same position. So, during the year, the Sun appears to shift its position slowly against the backdrop of stars, as we view it from differing positions in space.

This means that the Sun appears to travel once round the celestial sphere during the course of a year, so describing a great circle as it goes. This great circle, the annual path of the Sun, is called the *ecliptic*. Remembering that the celestial equator is simply the projection of the Earth's equator on the celestial sphere and that the Earth's Equator is inclined at an angle of $23\frac{1}{2}°$ to the plane of the Earth's orbit about the Sun, it can be seen that the ecliptic is inclined at an angle of $23\frac{1}{2}°$ to the celestial equator (see Figure 1.19, *overleaf*).

The ecliptic cuts the celestial equator at two positions. On 21 or 22 December the Sun is at its furthest south from the celestial equator and afterwards moves northwards. On 21 or 22 March the Sun crosses the celestial equator at the point known as the vernal equinox (the zero point from which right ascension is measured). The Sun continues to move northwards as it progresses round the celestial sphere until 21 or 22 June, when it is then at its furthest point north of the celestial equator.

The Sun continues its path around the celestial sphere but it is now moving southwards, until it crosses the celestial equator, at the point known as the

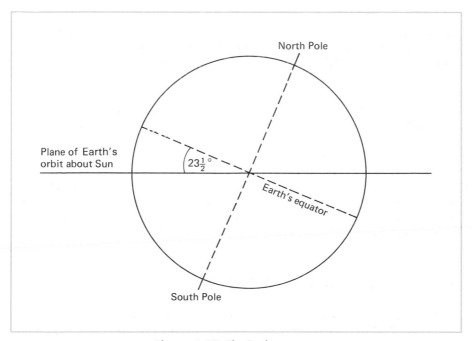

Figure 1.17 The Earth in space.

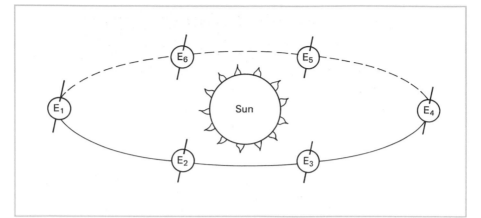

Figure 1.18 The Earth's orbit about the Sun.

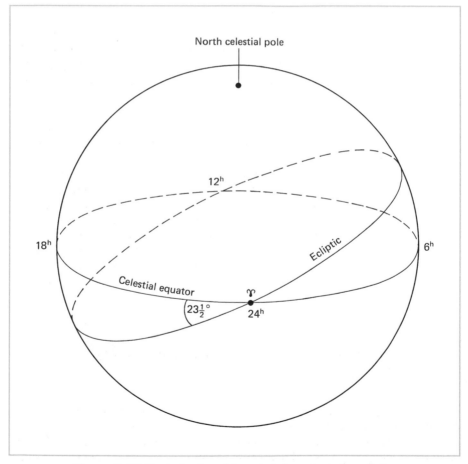

Figure 1.19 The inclination of the celestial equator to the ecliptic.

Table 1.1 The Sun's changing RA and declination

	Right ascension	Declination (degrees)
December 21/22	18 hours	$-23\frac{1}{2}$
March 21/22	24 hours	0
June 21/22	6 hours	$+23\frac{1}{2}$
September 21/22	12 hours	0

autumnal equinox, on 21 or 22 September. The Sun regains its original position far south of the celestial equator on 21 or 22 December. Thus the right ascension and declination of the Sun are continually changing over a yearly cycle (see Table 1.1).

1.12 The Seasons

When the Sun is at its most southerly declination the northern hemisphere of the Earth experiences the least number of hours of sunlight each day. Also, the Sun never attains a very high altitude as seen from this hemisphere. The result is that the northern hemisphere of the Earth is then experiencing winter. The time of maximum southerly declination of the Sun is known as the *winter solstice*, in the northern hemisphere. At this time the southern hemisphere of the Earth experiences the maximum number of hours of daylight per day, with the Sun reaching a high midday altitude.

So, northern winter occurs at the time of southern summer. The whole situation is reversed when the Sun is at its most northerly declination. We then have northern summer and southern winter. The time of maximum northerly declination of the Sun is known as the *summer solstice* for the northern hemisphere.

As far as the climate is concerned, spring and autumn are centred on the time of the equinoxes (though officially the equinoxes mark the **start** of spring and autumn – spring for the northern hemisphere beginning at the time of the vernal equinox). At these times the Sun is on the celestial equator and day and night are then of equal length.

1.13 The Moon and the Celestial Sphere

In the same way that the Sun is at the centre of the orbit of the Earth, so the Earth is, in turn, the centre of the orbit of the Moon. The mean Earth–Moon distance is 384 400 km, the Moon taking 27.3 days to go once round the Earth. This period is known as the *sidereal period* of the Moon. In this time the Moon will appear to go exactly once round the celestial sphere.

Since the Moon's orbital plane has a similar inclination (only 5° different) to that of the Earth's orbit about the Sun, the apparent path of the Moon across the celestial sphere closely follows that of the ecliptic. So it is that, over a period of 27.3 days, the Moon passes across twelve constellations, the same twelve constellations that the Sun takes 12 months to pass across. These are the constellations of the *zodiac*.

The Moon and Sun both travel in the same direction – eastwards – through the stars but the Sun travels very much more slowly. Hence, between successive alignments of the Moon and the Sun, the Moon has to travel a bit more than just once around the sky each orbit. Now, as we shall see in Chapter 6, a New Moon can only arise when the Moon and the Sun are very nearly in the same line of sight. The result is that the interval between successive New Moons is just a little bit longer than the Moon's sidereal period. It is actually 29.5 days and is known as the Moon's *synodic period*.

1.14 The Solar System and the Celestial Sphere

About 5000 million years ago the Sun and the planets condensed out of a cloud consisting mainly of hydrogen enriched with heavier elements spattered into space by old exploded stars. As the cloud contracted it swirled into a vortex and, on further contraction, conservation of angular momentum caused the rate of rotation to increase. Eventually a flattened disk was formed. It was from this disk that the central Sun condensed surrounded by rings of matter that gave rise to the planets.

As a remnant of this early period of creation the planets all orbit the Sun in very nearly the same plane, none of the planetary orbital planes being inclined at more than 7° from the mean. Pluto, on the edge of the system, is the exception with an orbital inclination of 17°.2 but Pluto is in many ways a maverick. It is a very small body and has a highly eccentric orbit. Thus, with the exception of Pluto, all the planets of the Solar System appear to us to move along the same band of sky (the zodiac) as they proceed along their orbits around the Sun.

1.15 Stellar Motions

It would be fanciful to imagine that the stars are rigidly fixed in space with respect to each other. Indeed, the stars do have their own individual motions in space. However, in terms of the vast distances separating them, their individual motions are not enough to make any noticeable changes to the shapes of the constellations for thousands of years. The stars today look pretty much the same to us as they did to our ancestors. The displacement of the stars on the celestial sphere due to their own motions through space are known as *proper motions*. The values and directions of the proper motions naturally differ, the largest being of the order of 10 seconds of arc per year. Others have proper motions a great deal smaller.

1.16 Precession

The Earth is not a perfect sphere but bulges outwards somewhat at its equatorial regions. This is due to the centripetal force as the Earth turns on its axis. The Sun and the Moon act gravitationally on this bulge and cause the Earth to "wobble" in its orbit in the same way that a child's spinning-top wobbles as it is slowing down. In the case of the spinning-top the time for one gyration is little more than a couple of seconds. In the case of the Earth a period of 25 800 years elapses while the axis of the Earth *precesses* once.

Thus the position of the celestial poles is constantly, though very slowly, changing as the spin axis of the Earth precesses. At present the position of the north celestial pole lies close to the star Polaris but in 12 000 years time the bright, blue, star Vega in the constellation of Lyra will be in position as the pole star. 25 800 years from now precession will have shifted the position of the north celestial pole in a large circle, of diameter 47°, and it will be once again marked by Polaris.

Questions

1 Draw a diagram of the celestial sphere labelling the celestial poles, the celestial equator, and the observer's horizon. Show the Earth at the centre and draw the positions of the terrestrial poles and the equator, being careful to show their relation to the celestial sphere.

2 Define a *great circle*. Explain the term *meridian* in the context of (a) the Earth and (b) the celestial sphere. Explain fully how the coordinate system for navigation on the Earth's surface has been set up. Your answer should include the significance of the Greenwich Meridian.

3 Explain fully the coordinate system used by astronomers for celestial objects, including such points as: positive and negative declinations, the Vernal Equinox, the rotation of the Earth, sidereal time, etc.

4 What causes the stars to move across the sky? Why is the apparent motion of the Sun different from that of the fixed patterns of stars? Explain your answers. Give the common name for these patterns of stars. What is the apparent path of the Sun through the stars called? What does this, in fact, represent?

5 Define *zenith* and *nadir*. Explain the terms *altitude* and *azimuth*, showing how they are used in setting up an alternative to the equatorial coordinate system. What are the major advantages of each system?

6 Draw a diagram that explains the relationship between latitude and the altitude of the pole star from a given location. Include in this diagram

the observer's horizon and the direction of the zenith. What is the greatest altitude of the celestial equator as viewed from a latitude of 35° N and in which compass direction does it achieve its greatest altitude?

7 Why will the altitude of a star vary during the night as viewed from a mid-northerly latitude? Where on Earth would the star be visible above the horizon all the time and have a constant altitude? Explain the terms *upper culmination* and *lower culmination* and in the process define the term the *observer's meridian*.

8 What is the altitude of Polaris as seen from an observatory that has a latitude of 50° N? What is the maximum altitude of a star, as seen from this observatory, that has a declination of (a) +10° and (b) –15°? What is the declination of a star that will *just* not rise above the horizon from this latitude? What is a *circumpolar star* and what is its minimum declination in order to be circumpolar as seen from this latitude?

9 Explain, with the help of diagrams, the inclination of the ecliptic to the celestial equator and explain how this gives rise to the right ascension and declination of the Sun changing throughout the year.

10 Explain the terms: *summer solstice*, *winter solstice*, *vernal equinox* and *autumnal equinox*. Explain how the seasons arise on the Earth. What is the *zodiac*? Why might the RA and the declination of a star vary over a period of a few centuries?

11 You receive a letter from your Great Aunt Matilda, who lives in East London. In it she says:

> I was very interested to learn that you are studying the stars. Perhaps you can explain some things to me, dear?
>
> I know that we go round the Sun and this is what causes the stars to move across the sky – but why does the Pole Star stay still all the time?
>
> When I went to Australia, all the stars were different and I couldn't see the Pole Star at all. Why was this?
>
> Also, where do the Pole Star and all the other stars go in the daytime?
>
> I've wanted to know these things since I was a girl. If you could explain them to me I will give you a nice big hug when I visit you next!

Clearly Great Aunt Matilda is confused about some matters as well as having some legitimate questions. Write her a reply which deals with her misconceptions and answers all her questions. You can include diagrams to help you with your explanations.

12 You receive another letter from Great Aunt Matilda (see the previous question). She thanks you for explaining things so clearly to her and then goes on to write:

> I have another question for you: I know that we have summer when the Earth gets close to the Sun and winter when we are far away from it – but

why was it so terribly hot in Australia when I went last Christmas? I got many strange looks because I was dressed in my heavy winter woollies but I refused to take them off. After all, it was winter wasn't it, and I always wear my woollies in the winter. Someone there said it was the hottest time of the year but that is silly and I told them so, too! How could that be at Christmas – they would not believe me when I tried to tell them. They are a strange lot, those Australians!

Oh dear! Obviously dear old Great Aunt Matilda needs another letter to straighten things out for her. Write a reply which will sort out her misconceptions and explain the truth to her. As before, you can include diagrams to aid your explanations.

Chapter 2

The Earth and Time

THE ancients considered the Earth to be flat and located at the centre of the Universe, with the sky a crystal dome on which the stars were fixed. It was around 2000 BC that Aristotle realised that the Earth is a globe. There is clear evidence to support this view. For instance, there is the differing altitude of the pole star as seen from different locations and the fact that the most southerly of the constellations are only rendered visible by travelling southwards (the converse is true for people living in the southern hemisphere). More commonly known is the phenomenon of the lower parts of a ship which is travelling away from the shore disappearing from view while its sails are still visible above the horizon. The sails are then seen to 'sink' below the horizon as the ship further recedes.

2.1 The Shape and Size of The Earth

The next major advance came in 280 BC when Eratosthenes measured the circumference of the Earth. He was in charge of the vast library of scrolls at Alexandria. He knew from one of these scrolls that on noon at midsummer's day in Syene (now called Aswan) the Sun cast no shadows of vertical objects. In other words, the Sun was then directly overhead at Syene. By direct measurement Eratosthenes found that, at the same instant, the Sun was $7\frac{1}{2}°$ from the zenith as viewed from Alexandria. Now, $7\frac{1}{2}°$ is roughly one-fiftieth part of a circle and so he reasoned that the distance of Alexandria from Syene was one-fiftieth of the total circumference of the Earth.

In this way he calculated the circumference of the Earth to be 240 000 stadia, which converts to a value of about 40 000 km, remarkably near the truth. It was not until AD 1492 that Christopher Columbus finally dispelled doubts about the true shape of the Earth when he partially circumnavigated it! The principle of Eratosthenes's method is illustrated in Fig. 2.1.

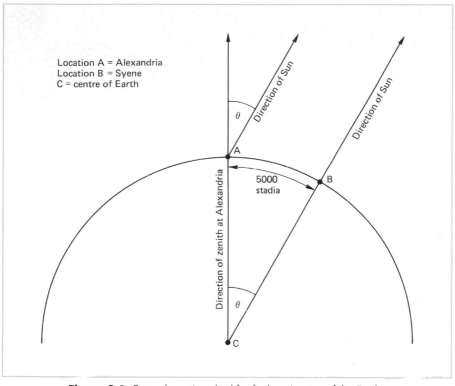

Figure 2.1 Eratosthenes' method for finding the size of the Earth.

2.2 The Earth in the Solar System

Running counter to the religious dogma of the time, ideas that the Earth orbited the Sun were stifled and the Greek scholars stayed with the idea of the central Earth, or "geocentric theory". It can be argued that this held up the progress of astronomy for at least a thousand years.

In 1543 Copernicus had his work *De Revolutionibus Orbium Coelestium (Concerning the Revolutions of the Celestial Bodies)* published while he was on his deathbed. He had completed his work some years before but withheld publication for fear of reprisals from the Church authorities. Copernicus proposed the "heliocentric theory" in his work, a scheme which fitted the observed motions of the planets much better than did the geocentric theory. The reprisals of the Church were indeed extreme, as Copernicus had so wisely guessed. In 1600 Giordano Bruno was burnt at the stake, in part for his teachings of the Copernican, rather than the Ptolemaic, view of the structure of the heavens.

It was left to Galileo (between 1609 and 1611) to provide practical proof for the heliocentric model. With the newly invented telescope he discovered mountains and craters on the Moon and sunspots, which shook the view that the heavens were perfect and unchanging. He also discovered that the milky band of radiance that crossed the starry sky – the Milky Way – was composed of innumerable faint stars. Worst of all, as far as the Church was concerned, Galileo discovered that Jupiter was attended by four moons that seemed to orbit the planet. This indicated that the Earth was **not** the centre of all things.

Further, he saw that Venus showed phases and displayed variations in apparent size in a way which was simply impossible if the Ptolemaic scheme was correct. The Church authorities tried to suppress the spread of the knowledge of Galileo's discoveries, though with only partial success. Galileo was forced to recant his discoveries and he lived out his last years under virtual house-arrest. However, it was too late. The truth was out and with it the main hurdle was removed and the science of astronomy was free to progress and prosper.

Today we know that the Earth is a rocky globe, 12 800 km in diameter, orbiting the Sun at a mean distance of 150 million km, taking one year to complete one orbit. We also know that the Earth is not quite a perfect sphere but is rather an oblate spheroid. The Earth's polar diameter is 43 km less than its equatorial diameter. The difference is caused by the forces arising from the spinning motion of the Earth on its axis.

The Earth is but one of nine major planets that orbit the Sun and they all rely on the Sun as a provider of light and heat. All the planets go round the Sun in the same direction and all (bar one maveric) keep close to the same plane as they orbit. The mean orbital radii of the planets and their orbital periods are shown in Fig. 2.2. The orbital period of a planet is referred to as its *sidereal period*. Hence the sidereal period of the Earth is one year.

2.3 The Structure of the Earth

The mass of the Earth is 6 million million million million kg. Dividing the Earth's mass by its volume gives it a mean density of 5520 kg/m^3. The rocks near the surface of the Earth have densities ranging about 2500 kg/m^3 and so it is obvious that the Earth is not uniform in density. In fact, the density increases towards the centre. This is because the Earth is chemically differentiated. By this it is meant that when the Earth was formed the lighter and more volatile substances rose to the surface while the heavier substances sank towards the Earth's centre.

Seismographic studies have revealed that three distinct zones exist within the Earth. These are the *core*, the *mantle* and the *crust*. The core extends from the Earth's centre and out to a radius of 3500 km and consists of two parts: a fluid outer core and a central solid core. The solid component is thought to be about 2500 km in diameter. The core is composed of iron, nickel and cobalt.

Over the core, out to a radius of about 6370 km, lies the mantle. It is mainly composed of silicates and is rather plastic in nature. Overlying the mantle, with

Figure 2.2 a The orbits of the inner planets. **b** The orbits of the outer planets.

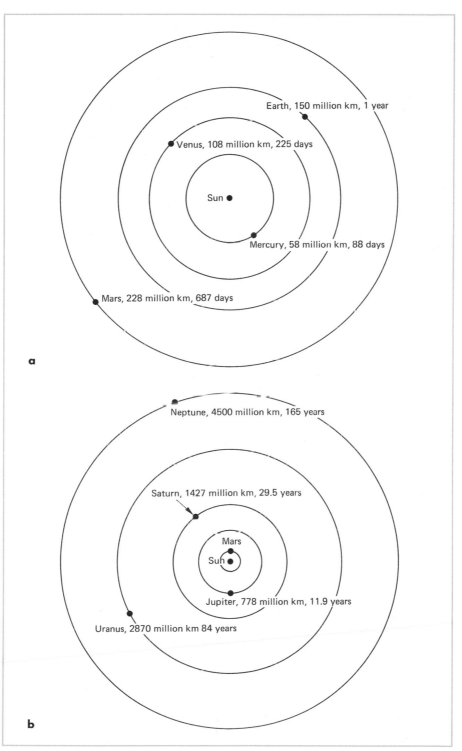

a

b

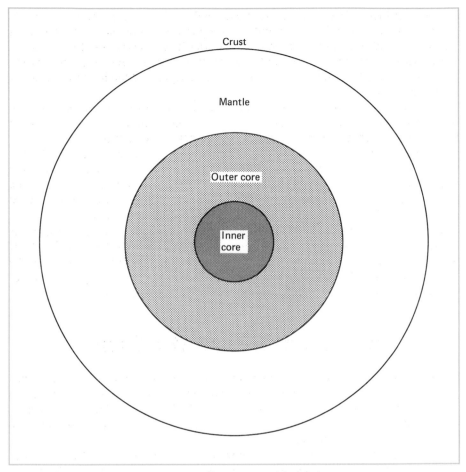

Figure 2.3 The structure of the Earth.

an average thickness of about 33 km, is the crust upon which we live. Figure 2.3 diagrammatically shows the structure of the Earth. Radioactive decay, together with the heat left over from the formation of the Earth, causes the temperature to increase towards its centre. The temperature at the Earth's core is thought to reach about 5000 °C.

2.4 The Surface Conditions on the Earth

The surface of the Earth is mainly composed of basaltic, sedimentary and volcanic rocks (a basalt is a fine-grained type of volcanic rock mainly composed of the minerals pyroxene and plagioclase). A most notable feature of this planet is that it is geologically active. Indeed, the very crust of the Earth resembles a huge broken eggshell with each of the pieces of shell floating on the mantle below it. These segments are called *plates* and the study of their movements is called *plate tectonics*; hundreds of millions of years ago the land masses over the surface of the Earth were grouped into one large continent since when tectonic activity, more commonly known as *continental drift*, has separated the land masses as we see them today.

Enormous forces are pent up at the boundaries between the crustal plates. Along these *fault lines* earthquakes are propagated when one plate slips against (or under, or over) another. Volcanoes are also active along fault lines.

About two-thirds of the Earth's surface is covered in water. It is thought that life initially developed in the seas and it is certainly true that the earliest fossil remains of life on this planet are of primitive marine creatures. Indeed, the chalk and limestone deposits that are so obvious along our shores are the remains of sea creatures that lived and died over a hundred million years ago.

The range of temperature that is experienced on the surface of the Earth is moderated by the presence of the atmosphere, which has a blanketing effect. Heat is retained during the night and the surface is shielded from the harshest of the Sun's rays during the daytime. The temperatures on the Earth range from about −40 °C at the Earth's frozen poles to around +40 °C at the Equator.

2.5 The Atmosphere of the Earth

The sea level pressure of the Earth's atmosphere is 101 325 N/m^2 (newtons per square metre). The air is composed of a mixture of different gases. Its composition by volume is nitrogen (78%), oxygen (21%), carbon dioxide and argon making up most of the remaining 1%. The lower portions of the atmosphere are also quite rich in water vapour.

There are traces of atmosphere hundreds of kilometres up but most of the mass of the air is concentrated in the first 10 km. As Fig. 2.4 (*overleaf*) shows, the atmosphere can be divided up into distinct layers. These are marked by temperature variations. The lowest region is the *troposphere*, which is the region of weather. All the cyclones and storms – as well as our unreliable British weather – occur in this region. As the height above the ground increases so the temperature drops (as does the pressure) until a region 12 km up, the *tropopause*, where the temperature levels off at about −70 °C. Cirrus clouds can extend up to a height of about 10 km.

Beyond the troposphere is the *stratosphere*, where the temperature begins to rise slowly until a region known as the *stratopause*, where the temperature levels off at about −40 °C. The *stratopause* lies at a height of 40 km and beyond this lies the *mesosphere*. In this region the temperature falls rapidly to a value of about −90 °C at an altitude of 80 km, the *mesopause*. Above the mesopause lies the *thermosphere* where the temperature rapidly increases to 1000 °C and then levels off at 600 km in a region known as the *thermopause*. Finally, above the thermopause lies the *exosphere*, which consists mainly of hydrogen and extends to a height of about 5000 km where it merges with interplanetary space.

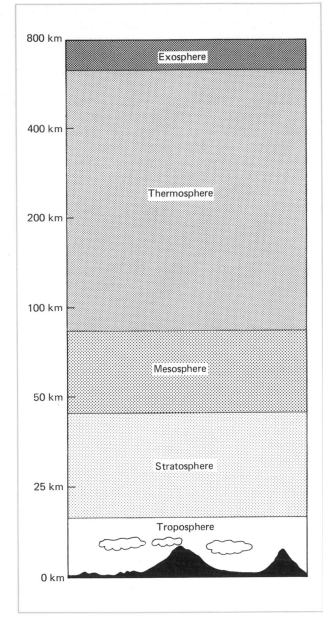

Figure 2.4 The Earth's atmosphere.

2.6 The Ionosphere

As well as light and heat, the Sun emits copious amounts of short-wave radiations in the ultraviolet and X-ray regions of the spectrum (more about this in Chapter 15). These radiations have the effect of ionising the rarefied gases in the upper atmosphere. The *ionosphere* is usually taken to exist between 80 and

300 km above the Earth's surface, as a series of ionised regions and layers, though the regions above this are also highly ionised.

In 1901 Marconi found that radio waves could be reflected from the sky, thus allowing transmissions over very great distances despite the curvature of the Earth. We now know that it was the ionosphere that reflected the radio waves – a fact invaluable to modern communications. The ionosphere is rich in ozone, formed by the ionisation of oxygen. This ozone is vital to the survival of life on this planet in that it absorbs much of the harmful short-wave radiation sent to us from the Sun. It is in the ionosphere that the aurorae are produced, but we must defer a discussion of this phenomenon until we have considered the Earth's magnetic field.

2.7 The Terrestrial Magnetosphere

It is common knowledge that the Earth possesses a magnetic field. It has been used as an aid to navigation for a great many years. The Earth's field is *dipolar* in form, like that of a simple bar magnet (Fig. 2.5). The diagram shows that the Earth's rotational and magnetic axes do not coincide. At present they are inclined at $11\frac{1}{2}°$ to each other.

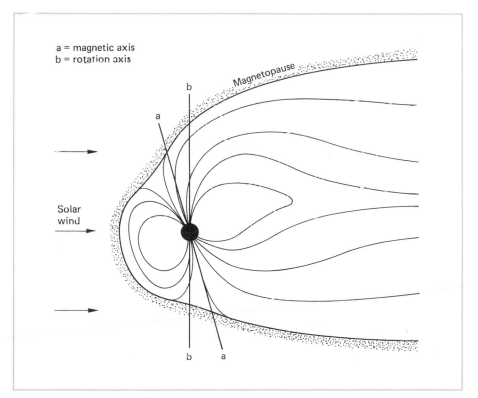

Figure 2.5 The Earth's magnetic field.

The positions of the magnetic poles are not fixed but appear to be moving, very slowly, in a very roughly circular path. At present the north magnetic pole is moving westwards by about $0°.05$ per year. In addition, the magnetic field is not constant in strength. It is currently decreasing at the rate of 1% per 10 000 years. The Earth's magnetic field has even been known to reverse its polarity. This has happened several times in the past few million years, as revealed by studies of ferromagnetic rocks near the surface.

We think that the terrestrial magnetic field is generated by a natural dynamo effect. The material of the Earth's core conducts electricity. Its churning motions (due to convection of the hot fluid and the spinning motion of the Earth) cause electric currents. These electric currents then generate the magnetic field.

Another fact, illustrated in Fig. 2.5, is that the Earth's magnetic field, or *magnetosphere*, is somewhat distorted. The reasons for this are twofold. One is that the Sun has a very powerful and extensive magnetic field that extends outwards into the Solar System, interacting with the Earth's field. The other reason is that the Sun emits a stream of electrified particles that blow through the Solar System like a breeze. The fact that these particles are electrically charged gives them the property of being able to distort the Earth's magnetosphere.

The Earth's magnetic field lines are flattened in the direction of the Sun and are extended in the antisolar direction. Near the Earth its own magnetic field is strong and so it is only slightly distorted by solar interactions. However, the field strength decreases in accordance with an inverse cube law and so the field strength at large distances from the Earth is low and the shape of the field is distorted much more.

At about 60 000 km from the Earth the field lines no longer rejoin at the Earth's magnetic poles, this region being termed the *magnetopause*. Beyond the magnetopause lies a region where a large fraction of the particles from the Sun are deviated in their courses and pass round the Earth and out into the Solar System under the repulsive effect of the Earth's magnetic field. This region is known as the *magnetosheath*. However, some rather more energetic solar particles are trapped lower in the Earth's magnetosphere and this brings us on to the Van Allen radiation belts.

2.8 The Van Allen Radiation Zones

James Van Allen studied the results of measurements of radiation intensity at various heights using rockets and satellites sent aloft in 1958. These measurements indicated that two toroidal zones of very high energy particles exist high above the Earth. As illustrated in Fig. 2.6, these are regions where the Earth's magnetic field has trapped high energy particles from the Sun. The inner torus is filled with protons of energies ranging about 50 MeV (mega electronvolts) and lies at an average height of about 4000 km above the Earth's surface. The outer torus, which lies at an average height of about 17 000 km, is filled with electrons with energies ranging around 30 MeV. These energetic particles loop from end to end within the zones.

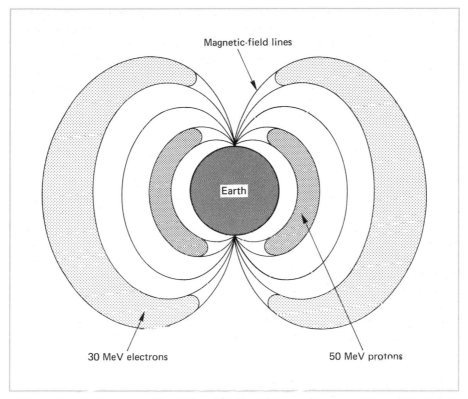

Figure 2.6 The Van Allen zones.

2.9 The Aurorae

Aurorae have been known since ancient times. They take the form of beautifully coloured glows in the night sky. They are most often seen from far northerly and far southerly latitudes. In fact, they are visible on most nights from countries like Greenland and Alaska. They are least often seen from the equatorial regions of this planet.

The colours of the displays are chiefly reds and greens, though bluish tints are sometimes seen. The display may take the form of pulsating, coloured glows in large, diffuse, patches of the sky, or perhaps just steady reddish glows. A spectacular display may consist of ribbons and streamers of colour that pulsate and change their shapes in a period of a few minutes. From mid-northern latitudes aurorae can sometimes be seen by looking towards the northern horizon, though town lights will completely swamp the effect. The peoples of the mid-southern hemisphere should look towards their southern horizon. The aurora concentrated toward the north pole is referred to as the *aurora borealis*, while that concentrated towards the south pole is called the *aurora australis*.

The aurorae are caused by the Van Allen radiation zones becoming over-loaded with particles so that some of them spill out and spiral down the magnetic field lines toward the Earth. The particles ploughing through the ionosphere consititute electrical currents that give rise to the coloured glows in the same manner as in the gases of the low-pressure discharge lamps so commonly used today.

It is interesting to note that, since the particles in the Van Allen Zones are rapidly oscillating in a north–south direction from one extent of each zone to the other, very similar displays are set up in opposite hemispheres of the Earth at the same instant.

The auroral displays tend to occur with variable frequency and intensity, since the Sun does not send out its breeze of electrified particles at a constant rate. There is much more about the Sun later in this book, but suffice it to say here that the Sun's activity swells and dies away with a period of about eleven years. At times of high solar activity auroral displays are at their most frequent and most spectacular.

I well remember the brilliant aurora on the night of 13 March 1989. At dusk that evening I was busy with one of the telescopes of the Royal Greenwich Observatory at their former site at Herstmonceux. At one point early on I left the dome and noticed a peculiar glow along the northern horizon, like a row of searchlight beams fanning upwards. As I watched I could see the glow wavering slightly and changing in brightness.

I had observed the Sun earlier that day and noted its high activity (it was close to the time of maximum activity in its cycle). I realised what was happening and ran to alert the only other observer on site at that time. We climbed onto the roof of an adjoining building between two domes and watched and photographed as an amazing auroral display unfolded. At times vast arches of blue/green light pulsed low down in the north, with upward shafts and wavering curtains of light becoming a rich vibrant red extending to the zenith.

The display continued. At one point the whole sky became a vivid red colour and the scene around us became quite unearthly. Later a white patch formed at the zenith, which broke into moving ripples at the same time as tickertape-like darts of light formed in the south. The telephone lines to the observatory became jammed and that aurora became an international news story. It was an awesome experience and yet it is rare to see the aurora at all from a latitude of 51° N!

2.10 The Need for Time Measurement

The division and measurement of time is tremendously important to mankind. However, most people are ignorant of the fact that time measurement and the classification of its units are based upon astronomical events. Indeed, the chief work of the early astronomical observatories was to mark and measure the passage of time. Their dedication to the purpose of studying the physical natures of the heavenly bodies is relatively recent!

2.11 Calendars

Calendars have been constructed since antiquity, each new one due to some refinement that allowed it to follow the passing of the seasons more faithfully without becoming out of step with them. However, most calendars were based on the interval known as the year. The year is the time taken for the Sun to appear to go once round the sky, with reference to the "fixed stars". We now know that the Earth revolves about the Sun, taking one year to complete its orbit. A major problem arose with the old calendars because of the fact that the Earth does not go round the Sun in a whole number of days, but in fact takes 365.256 36 days.

Julius Caesar masterminded the modification of adding one extra day, to a 365 day year, to every fourth year. The additional day is added in February, which is the shortest month of the year. In order to determine whether a particular year is a *leap year* or not, simply divide by four. If there is no remainder it is a leap year and February has 29 days, rather than its usual 28. Unfortunately, this scheme was not quite accurate enough and in the modern calendar, called the *Gregorian calendar* after Pope Gregory XIII, a further modification is made in that the century years (1900, 2000, etc.) are only leap years if they are exactly divisible by 400. Thus 1900 was not a leap year but 2000 will be. Obviously, a small accumulative error will still be present in any artificial system such as this, but since this error will take more than 3000 years to amount to one day, it is ignored.

2.12 The Solar Day

Everyday life requires a time system geared to the passage of day and night, so this system must be linked to the apparent motions of the Sun. Successive transits of the Sun across the observer's meridian defines the length of the day. However, the fact that the Earth's orbit is elliptical and that the ecliptic is inclined to the celestial equator means that the rate of change of right ascension of the Sun varies during the year. This, in turn, means that the interval of time between successive solar transits of the meridian varies a little during the course of the year.

So, in order to set up a scale of time measurement, we cannot use the real Sun, referred to as the *apparent Sun*, and astronomers have devised a fictitious *mean Sun*. The mean Sun sometimes lags behind and sometimes moves in front of the apparent Sun, though the difference is never large.

The interval between successive transits of the mean Sun across the observer's meridian defines the length of the *mean solar day*. The mean solar day is then divided up into 24 hours of *Mean time*. The correction to *apparent solar time* to give mean solar time is very straightforward to apply and is known as the *equation of time*. A discussion of the equation of time is deferred until later in this chapter.

Imagine that an observer is situated at a certain longitude on the Earth. Further suppose that he sees the Sun on his meridian at a certain time. It should be clear

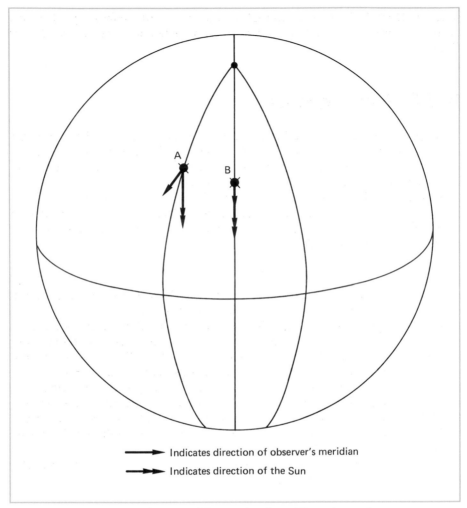

Indicates direction of observer's meridian

Indicates direction of the Sun

Figure 2.7 The Sun and the meridian from different longitudes.

that another observer, at another longitude, would **not** see the Sun on **his** meridian at the same time (see Fig. 2.7). In other words, midday will occur at different times for different longitudes, so each longitude will have its very own time-scale! At one time communities did indeed have their own time-scales but increased communications and travel necessitated some procedure of standardisation.

2.13 Universal Time

Due to its naval importance, the Royal Observatory at Greenwich adopted itself as the timekeeping centre of the world and defined *Greenwich Mean Time* such that when the mean Sun was on the observer's meridian as seen from Greenwich the time was $12^h\ 00^m\ 00^s$ GMT. In fact GMT became so universally adopted for the purposes of astronomy and navigation that it became known as *Universal Time* (UT).

2.14 Time Zones

During the course of a day the Sun travels an angular distance of 360° around the sky. The Sun appears to move through an angular distance of 15° every hour. Alternatively, for every 15° of longitude difference in a westerly direction the Sun is an hour later in reaching the observer's meridian.

In order that time is related to the hours of sunlight, the Earth is divided into 15° bands of longitude. Each band has its own *time zone* which is related to Universal Time by the addition or subtraction of whole hours of time. Thus we define *zone mean time* (ZMT) as follows:

$$ZMT = UT + L/15$$

where L = the central longitude of the time zone and is positive for locations east of the Greenwich meridian and negative for those to the west.

In practice, political and geographic factors cause deviations of the boundaries of the time zones. A further complication to our scheme is the use in the UK of British Summer Time (BST) when an hour is added to GMT so that the inhabitants of the British Isles may enjoy more hours of sunlight in the warm summer evenings without changing timetables and affecting commerce. Daylight Saving Time and even Double Daylight Saving Time are similar and are used elsewhere. In addition, at a longitude of 180°, a time zone at 24^h GMT is adjacent to one at 0^h GMT, and so here occurs the *International Date Line*; when this line is crossed the date changes by one whole day.

Fortunately, astronomers have little to do with such complications and refer most of their calculations to sidereal time.

2.15 Hour Angle

We have already seen that a star on the celestial sphere can be thought of as lying on a celestial meridian (a great circle joining the celestial poles). It is important to understand that this particular great circle is fixed with respect to the stars and is thus moving, at the diurnal rate, with respect to the observer's meridian. This great circle is known as the *hour circle* of the star considered.

The angle that the hour circle makes with the observer's meridian, measured westwards from the meridian in units of time, is called the *hour angle* of the star. The hour angle measured from the observer's meridian (at whatever longitude his location might be) is known as the *local hour angle* (LHA) of the celestial body being considered, while the hour angle of the celestial body as measured from the observer's meridian at Greenwich is termed the *Greenwich hour angle* (GHA). Using the idea of hour angle we can now go on to consider sidereal time.

2.16 The Sidereal Day

Just as the interval between successive transits of the Sun is equal to the solar day, so the interval between successive transits of a given star is the *sidereal day*. Due to the Earth's orbit about the Sun, the Sun appears to move eastwards through the stars, completing one 360° circuit in one year. This amounts to slightly under 1° per day. Thus the solar day is about 4 minutes longer than the sidereal day. In fact, the sidereal day has a length of $23^h 56^m 04^s$, as measured in solar time.

Once again, the vernal equinox is used as a marker. When the this point transits the observer's meridian the *local sidereal time* (LST) is $00^h 00^m 00^s$. The local sidereal time is then always equal to the local hour angle of the vernal equinox.

Similarly, the *Greenwich sidereal time* (GST) is equal to the Greenwich hour angle of the vernal equinox. Note that, since the sidereal and solar days are each divided up into 24 equal parts, or hours, the sidereal hour and the solar hour are not quite equal in length.

2.17 A Fundamental Equation

The right ascension of a celestial body is the difference in time between meridian transits of the vernal equinox and the considered body. Sidereal time is given by the hour angle of the vernal equinox, and therefore the hour angle of a celestial body is equal to the difference between the sidereal time and its right ascension. This can be put into the form of an equation:

$$HA = ST - \alpha$$

This equation can be rearranged to find the sidereal time, knowing the values of right ascension and hour angle of a given celestial object:

$$ST = HA + \alpha$$

LHA, GHA, LST and GST can also be distinguished. Thus:

$$GHA = GST - \alpha$$
$$GST = GHA + \alpha$$
$$LHA = LST - \alpha$$
$$LST = LHA + \alpha$$

The usefulness of the basic equation and its rearranged forms is illustrated in the following example questions:

Example 1

A star of right ascension $2^h 30^m 20^s$ has an hour angle of $4^h 20^m 10^s$ at a certain instant. What is the local sidereal time at that instant?

Using   LST = LHA + α
               LST = 4^h 20^m 10^s + 2^h 30^m 20^s
               LST = 6^h 50^m 30^s

Example 2

A star of right ascension 16^h 10^m 15^s has an hour angle of 19^h 30^m 00^s as measured from the observer's meridian at a certain moment. What is the local sidereal time at that moment and what is the angular distance of the star from the meridian, measured in degrees? Is the star east or west of the meridian?

Using   LST = LHA + α
               LST = 19^h 30^m 00^s + 16^h 10^m 15^s
               LST = 35^h 40^m 15^s

This is greater than 24^h so we can subtract this amount to give the LST:

               LST = 35^h 40^m 15^s–24^h 00^m 00^s
               LST = 11^h 40^m 15^s

Since hour angle increases westwards, the star is 19^h 30^m 00^s west of the meridian. Every hour corresponds to an angle of 15°, so the star is 292°.5 west of the meridian. This is better expressed by saying the star is 67°.5 **east** of the meridian (360°.0 − 292°.5 = 67°.5).

Example 3

The star Arcturus ($\alpha = 14^h$ 14^m 36^s, $\delta = +$ 19° 18′ 05″) was observed to transit the meridian (at upper culmination) at Mount Wilson Observatory (latitude = + 34° 13′ 00″, longitude = 118° 04′ 00″ W) at a certain time and date. (a) What was the local sidereal time at the instant of transit? (b) What was the altitude of Arcturus at the time of transit? (c) What was the Greenwich sidereal time at the instant of transit?

(a) The local sidereal time at any instant is always equal to the right ascension of the observer's meridian, where it cuts the celestial equator at upper culmination. Since we know that the right ascension on the meridian at that instant was 14^h 14^m 36^s, it follows that the local sidereal time at the moment of transit was 14^h 14^m 36^s.

(b) The altitude at upper transit is the maximum altitude that a star can have from a given location and is equal to the colatitude of the location plus the declination of the star.
Altitude at transit = (90° − 34° 13′ 00″)
                    + 19° 18′ 05″
                    = 75° 05′ 05″

(c) Greenwich is 118° 04′ 00″ east of Mount Wilson. Now, each 15° eastwards corresponds to 1 hour more of sidereal time. Thus, dividing the longitude of Mount Wilson by 15 will tell us how far ahead Greenwich is in sidereal time. This is $7^h 52^m 16^s$. Thus:

$$GST = 14^h 14^m 35^s + 7^h 52^m 16^s$$
$$\underline{GST = 22^h 06^m 51^s}$$

2.18 The Measurement of Time

It should now be obvious that astronomical measurements form the basis of all timekeeping. Nowadays we use reference timepieces (for instance atomic clocks) and calibrate our various mechanical and electrical clocks to them. Of course, this was not always possible, and long ago all clocks had to refer to the motions of a particular celestial body – namely the Sun.

The simplest form of clock was the *shadow clock* which consisted of a pole erected perpendicular to the level ground. The Sun then cast a shadow onto a calibrated dial. As the Sun moved during the day, so the shadow moved over the numbers, indicating the local solar time. Unfortunately, the rate at which the shadow swept across the dial varied with the altitude of the Sun, so making the dial non-linear. The rate was also dependent on the declination of the Sun, since this alters the altitude. Thus a dial graduated for a given month would become gradually more inaccurate as the seasons progressed.

One way out of this problem would be to incline the pole so that it was aligned in a direction parallel to the Earth's rotation axis. In other words, it would then be pointed to the celestial pole. In this condition, the pole qualifies to be called a *style*. If the dial were set perpendicular to the style, the Sun's shadow cast by the style would sweep across it at a uniform rate during the day. The dial could then be graduated with equally spaced markings. The dial would also remain calibrated throughout the entire year. However, the snag with this arrangement is that from any one hemisphere of the Earth the Sun is only above the celestial equator for six months of the year. Hence the top of the dial will only be illuminated by sunlight for those six months.

The *sundial* utilises a horizontal dial but retains the idea of having a style inclined to the celestial pole (see Fig. 2.8). The sundial is usable whenever the Sun is shining and at any time throughout the year, although the shadow does not sweep over the dial at a uniform rate throughout the day. However, once set up, it does not go out of adjustment during the yearly cycle of seasons.

Two corrections have to be applied to the reading of a sundial in order to obtain the correct time. First, the sundial reads the local solar time. In order to correct to GMT allowance must be made for the longitude difference between the location of the sundial and the Greenwich meridian. The correction is to **add** 4 minutes for every degree **west** of the Greenwich Meridian (or substract the same amount if east of it). The second correction is to allow for the varying rate of travel of the Sun across the celestial sphere, during the course of the year, by

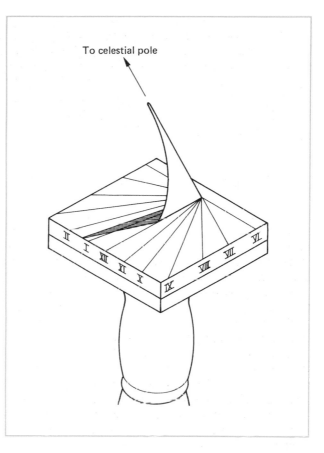

To celestial pole

Figure 2.8 The sundial.

applying the equation of time. This takes the form of a simple table of corrections which are applied using the formula:

Clock time = Sundial time – Equation of time

The table of corrections can be found in many publications, and never amounts to more than plus or minus $16\frac{1}{2}°$ minutes.

Questions

1   Discuss the evidence for the rotation of the Earth and its revolution about the Sun. Give the evidence for the shape of the Earth and describe the method by which its size was first measured. Draw a schematic diagram showing the position of the Earth in the Solar System and, on the diagram, give the orbital radii of the planets and their periods of revolution.

2   Describe as fully as you can the internal structure of the Earth and its major surface features and surface activity.

3   Give a detailed account of the structure and composition of the Earth's atmospheric mantle, being careful to explain the significance of the ionosphere.

4   Write an account of the Earth's magnetosphere, describing how it interacts with the Sun.

5   Write a short essay about the aurorae.

6   Explain how the modern calendar has evolved, outlining the difficulties with the older schemes. Briefly, why should mankind need a calendar at all?

7   Explain fully the terms *Greenwich sidereal time*, *Greenwich Mean Time*, *local sidereal time* and *local mean time*, showing how they are related to one another. Your answer should explicitly describe the concept of the *hour angle*.

8   Explain and show the need for *Zone Time*. Briefly outline the development of the sundial. What are the corrections that have to be applied to a sundial in order to convert the dial reading to GMT? Explain how the need for these corrections arises.

9   At 0^h UT on 1 June 1977 the Greenwich hour angle of the vernal equinox was $16^h 37^m 27^s$.

   (a)   What was the Greenwich sidereal time at 0^h UT?
   (b)   What was the local sidereal time at Flagstaff Observatory in Arizona at 2^h on this date?
         (Longitude of Flagstaff = 112° W, latitude = 35° 11′ N.)
   (c)   What is the maximum attainable altitude of the star Betelgeux, as seen from Flagstaff? (Coordinates of Betelgeux: $\alpha = 5^h 53^m$, $\gamma = +7° 24′$.)
   (d)   What is the local hour angle of Betelgeux from Flagstaff at 0^h UT on 1 June 1977, and at 2^h UT on the same date? (*Hint:* To give a precise answer you should allow for the slight difference in length of the sidereal and solar hours.)
   (e)   At what time GMT will Betelgeux transit the observer's meridian at Flagstaff?

10 At a certain instant at Meudon Observatory in France (longitude = 2° 14′ E) on 4 February a sundial reads $12^h 45^m$. What is the Greenwich Mean Time at this instant? (The Equation of Time for 4 February = -14^m.)

11 Find out about and write a paragraph on each of the following: (a) *leap seconds,* (b) *International Atomic Time.*

12 I see from my notebook that I observed a particular object, a star cluster in Hydra, on 16 February 1994. However, I had forgotten to write down the time I observed it, although I had noted that it was close to its greatest altitude. I have its coordinates written down as: $\alpha = 8^h\ 11^m.2$ and $\delta = -05°\ 38'$. My observing site is situated at a latitude of $51\frac{1}{2}°$ N and a longitude of $0°.375$ E. Out of curiosity I want to know the following:

(a) When did the star cluster reach its greatest altitude?
(b) What was that altitude, measured in degrees, above the horizon?
(c) In what compass direction was the star cluster from me, at that time?

(Looking up a table in an ephemeris, I find that $00^h\ 00^m$ UT on 17 February – midnight on 16 February – the Greenwich sidereal time is equal to $09^h\ 47^m$.)

13 (a) Which is longer: a solar day, or a sidereal day?

(b) To the nearest minute, how long is a sidereal day, measured in units of solar time?

(c) The star Rigel crosses the observer's meridian at $21^h\ 30^m$ UT from a particular observatory on 2 December. (i) What is the *local hour angle* of Rigel at that instant? (ii) At approximately what time (UT) will Rigel cross the meridian on 5 December?

Chapter 3

The Telescope

THE origins of the telescope are somewhat uncertain, though the Dutch spectacle maker Hans Lippershey is often credited with its invention. It is said that one day in 1608 he noticed that if he looked through a pair of spectacle lenses, held one in front of the other, he saw distant objects magnified. However, evidence has come to light in recent years which suggests that the telescope was invented some years earlier, though things are still far from clear-cut.

Galileo, among others, heard of the invention and, without any further information, designed a telescope for himself. This type of telescope uses a converging lens as the main light-collecting element and is known as a *refracting telescope*. A simple refractor consists of a lens of long focal length, the *object glass*, and a lens of short focal length, the *eyepiece*, to view the formed image.

In the form of telescope designed by Galileo the eyepiece is a planoconcave (diverging) lens. It has the advantage that it produces an image which is the correct way up (unlike Lippershey's refractor, which produces an upside-down image). However, the field of view in the Galilean instrument is very small and this is an important consideration if high magnifications are to be used.

The Galilean design was popular in the Victorian era, being used for opera glasses. However, modern telescopes use converging lenses (or a combination of lenses) for their eyepieces. Figures 3.1 and 3.2 show the designs of the simple refractor and the Galilean refractor.

3.1 Chromatic Aberration

One major problem with these early telescopes was that the images they gave were rather blurred and objects were seen surrounded by coloured fringes. We call this defect *chromatic aberration*. The trouble arose because the simple lenses used could not bring all the colours to a coincident focus (see Fig. 3.3).

The focal length of a lens divided by its diameter is known as its *focal ratio*, or f-number. A lens which has a diameter of 10 cm and a focal length of 1.2 m is an f/12 lens. The early experimenters found that the chromatic aberration in a lens could be reduced if the focal ratio, and hence the focal length, was made large. In the seventeenth century telescopes of longer and longer focal length

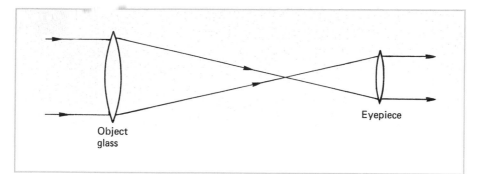

Figure 3.1 The simple refractor.

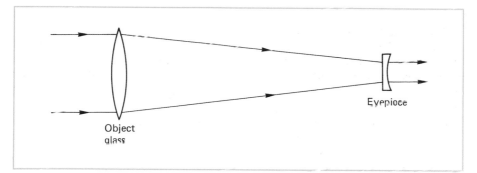

Figure 3.2 The Galilean refractor.

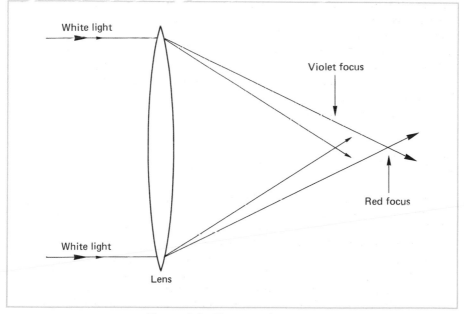

Figure 3.3 Chromatic aberration.

were constructed. These *aerial telescopes* did not have solid tubes. Instead, the object glasses were fixed to tall masts and taut ropes kept the eyepiece and the object glass in line. The eyepiece was mounted on a tall stand on the ground. It is said that the focal length of the longest telescope actually used was 210 feet (64 metres)! Needless to say, these unwieldy instruments were extremely difficult to operate and one marvels at the discoveries that arose from their use.

3.2 The Achromatic Objective

The late eighteenth century saw an invention that was to transform the refractor from an instrument of nearly unmanageable proportions to something more compact and of better optical quality. This was the achromatic object glass, principally developed by Chester Moor Hall and John Dollond.

Instead of a single lens, the refractor's object glass was constructed from two elements – a converging lens made of crown glass and a weaker diverging lens made of flint glass. The net result was that the object glass still had the necessary converging power, but the false colour effects inherent in each of the components tended to cancel out (see Fig. 3.4). Object glasses could now be made in much shorter focal lengths, typical focal ratios being in the range of f/10 to f/20.

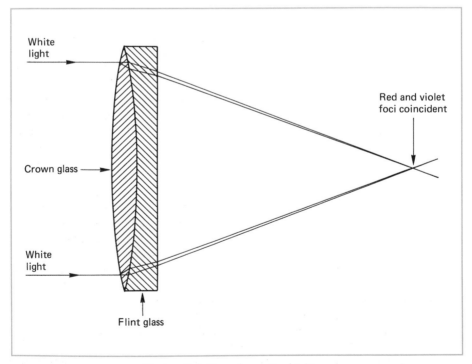

Figure 3.4 The achromatic objective.

Some of the larger aerial telescope object glasses had diameters around 8 inches (203 mm), but it was not until the beginning of the nineteenth century that flint glasses of good optical quality could be made of comparable size. This problem was solved in 1805 when Pierre Guinand devised a method of stirring molten glass, so giving rise to uniform flint blanks. (A "blank" is the circular slab of glass from which the lens is produced by grinding and polishing its surfaces to the correct shape.)

By the mid-nineteenth century opticians were making object glasses of around 15 inches (381 mm) diameter. This figure continued to increase, particularly due to the work of Sir Howard Grubb (in Ireland) and Alvan Clark & Sons (in the USA). Grubb's largest refractor has an *aperture* (astronomers call the diameter of a telescope's main light-collecting element the aperture of that telescope) of 28 inches (712 mm). This particular refractor, completed in 1894, is the seventh largest in the world. It is sited at Greenwich, in England.

The first and second largest refractors now in existence are both sited in the USA. The 36 inch (915 mm) of the Lick Observatory at Mount Hamilton, California, came into use in 1888. This was followed, in 1897, by the 40 inch (1.02 m) refractor of the Yerkes Observatory at Williams Bay, Winsconsin. The object glasses of both these telescopes were made by the Clarks, with mountings by Warner & Swasey.

3.3 Light Grasp and Resolving Power

Astronomers usually want the aperture of a telescope to be as large as possible for two very important reasons. The first is that a larger lens will collect more light than a small one, so allowing fainter objects to be seen. We say that the larger telescope has a greater *light grasp*. The second reason is that through a larger telescope one is better able to discern fine detail on a celestial body.

Any star, apart from our Sun, is so far away that it can appear only as a minute point of light. However, due to its wavelike nature (see Chapter 15) starlight passing through the aperture of the telescope experiences an effect we call *diffraction*. Even a perfect telescope objective will not produce an image of a star that is a perfect point. Instead, a *diffraction pattern*, consisting of a central disk surrounded by concentric rings of decreasing brightness, is formed.

The central disk (called the "Airy disk" after a former British Astronomer Royal) contains most of the light energy (84% for a perfect objective). A largescale representation of the diffraction pattern formed by a telescope, of a star, is shown in Fig. 3.5 (*overleaf*). If two stars are very close together in the sky they will only be seen as two, or *resolved*, if their diffraction patterns are not too large. Figure 3.6 (*overleaf*) shows the situation, for a given size of telescope, where the stars are non-resolvable, just resolvable and clearly resolvable.

It so happens that the sizes of the diffraction pattern produced in a telescope depend on its aperture. **Larger** apertures produce **smaller** diffraction patterns. Thus two stars that might not quite be resolvable as separate in a small telescope are clearly seen as two when using a larger one. The same is true for seeing fine detail on extended bodies (anything that does not appear to be a tiny point) such as the Moon or a planet.

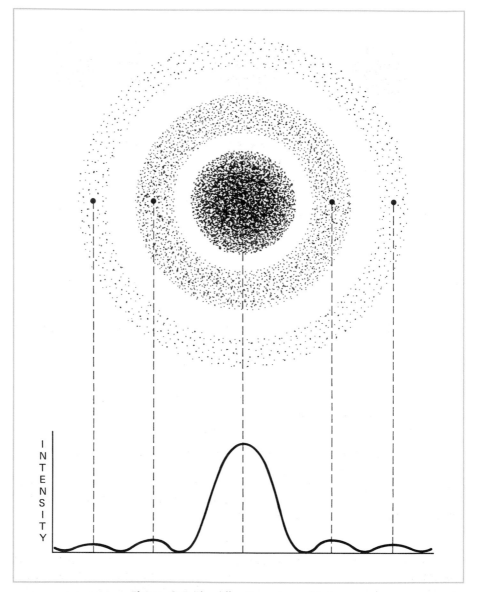

Figure 3.5 The diffraction pattern of a star.

Let us imagine that we are using a telescope to look at the surface of the Moon. We can think of the image as being composed of overlapping diffraction patterns – a bit like a mosaic made up of individual tiles. Obviously no detail finer than the size of one tile can be seen in the mosaic. A larger telescope produces smaller diffraction patterns. This is rather like having the mosaic made up of smaller tiles. Finer detail is rendered visible. Parts of the Moon might appear smooth and devoid of detail when seen through a 60 mm refractor. These areas might well show detail when seen through a 200 mm telescope.

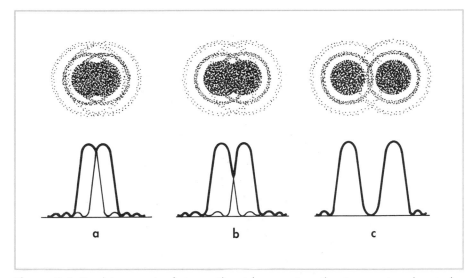

Figure 3.6 Resolving a pair of stars with a telescope. In **a** the stars are too close to be resolved. In **b** the stars are *just* resolvable and in **c** the stars are easily resolved.

A good guide to the *resolving power* of a telescope is given by the *Rayleigh limit*:

$$R = \frac{1.22\lambda}{D} \times 206\,265$$

where R = the angle of minimum resolution measured in seconds of arc,

λ = the wavelength of the light,

D = the diameter of the telescope's objective.

Both λ and D are measured in metres. How the minimum angle of resolution relates to the viewed image is explained in the next section.

Actually, an experienced observer can see details that are about 20% finer than that given by the Rayleigh limit and a more accurate prediction (applied to resolving pairs of stars) is given by the *Dawes limit*:

$$R = 4.56/D$$

where R is in arcseconds, as before, but D is in inches (1 inch = 25.4 mm). Note that this formula is only correct for visual wavelengths (taken as 5×10^{-7} m). Thus an $18\frac{1}{4}$ inch telescope should, in perfect conditions, be able to resolve stars that are separated by 0.25 arcsecond.

3.4 Eyepieces and Magnification

When using a telescope visually the choice of eyepiece is vital. The eyepiece may be just a simple lens, though the aberrations it produces are usually too great and the field of view too small, so a combination of lenses is used. It is the focal length of the eyepiece that determines the magnification obtained from a given telescope.

First, let us consider what we mean by magnification. In connection with telescopes we mean *angular magnification*. The Moon subtends roughly $\frac{1}{2}°$ as seen from the Earth. This is the size of a disk 1 cm across, seen from a distance of 1 m. If we view the Moon with a telescope giving a magnification of ×60, then it will appear to subtend an angle of 30°. Roughly speaking, this is the same as a disk 60 cm across seen at a distance of 1 m. In the telescope we have increased the *visual angle* of the Moon 60 times.

The magnification of a telescope is given by the formula:

$$\text{Magnification} = F_o/F_e$$

where F_o and F_e are the focal lengths of the objective and eyepiece respectively. In using the formula, the focal lengths of the eyepiece and objective have to be in the same units, but it does not matter what those units are (e.g. millimetres, metres, inches etc.).

Thus the magnification produced by any telescope depends on the focal lengths both of the objective and of the eyepiece. If the focal length of our objective lens is 1.5 m and we wish to obtain a magnification of ×60, then we must use an eyepiece of focal length 25 mm. It is usual to supply an astronomical telescope with a variety of eyepieces, so that the observer can choose the correct magnification to suit the object being viewed, the type of observation being carried out and the atmospheric conditions at the time of the observation.

3.5 The Effects of Magnification on the Image

The effects of magnification vary depending on the subject. For a star, magnification has virtually no effect (putting aside the bad effects of the Earth's atmosphere for a moment). If the telescope is a good one then a star will continue to appear as a point of light until very high magnifications are used. If a magnifying power as great as four times the aperture in millimetres (e.g. ×240 for a 60 mm aperture) is used then the diffraction pattern will become visible. Beyond this magnification the light from the star will become diluted as the diffraction pattern expands and so the star's image will become fainter.

Pairs of stars (known as double stars) at the limit of the resolution of the telescope can be split at a magnification roughly equal to the aperture in millimetres, for normally sighted people, though magnifications up to twice this value can be useful in allowing the stars to be separated without undue eyestrain. For double stars whose separations are at least several times the resolving limit of the telescope, increasing the magnification simply increases the apparent amount of black sky between the stars without altering the appearances of the stars themselves.

The effect that magnification has on extended objects is rather different. As the magnification is increased so the light from the considered body is more spread out. The image then looks fainter and the contrast of the details is reduced. For a faint extended object, such as a comet or a nebula, a low magnification must be used (ideally around one-sixth of the aperture in millimetres) since any increase will render the already dim object nearly invisible. Brighter comets and nebulae will stand more magnification. It is a quirk of optical geometry that too low a magnification (below about one-sixth of the aperture in millimetres) results in some of the precious light gathered by the objective not entering the observer's eye. Some of the light would thus be wasted.

Bright objects will stand more magnification. This is fortunate for the observer interested in seeing fine detail on the Moon and planets because here a magnification will be required that will render visible the finest detail that the telescope can show. However, it is also true that the image contrast is at a premium. A magnifying power roughly equal to the aperture in millimetres is usually preferable if the prevailing atmospheric conditions allow such a power to be used.

3.6 The Limitations Caused by the Earth's Atmosphere

The atmosphere of our Earth blocks off a large amount of the radiation arriving from space. Further, our atmosphere tends to adversely (as far as the astronomer is concerned) affect those radiations that do get through. The effects produced can be put into two catagories: *absorption* and *scintillation*. Both these effects are worse for celestial bodies observed at low altitudes (low angles above the horizon), because the body is then seen through the greatest thickness of air (see Fig. 3.7).

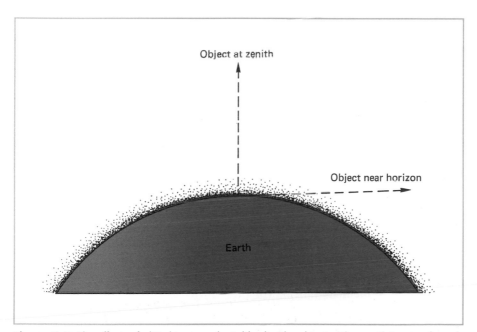

Figure 3.7 The effects of altitude on a celestial body. The object at the zenith is seen through much less of the Earth's dirty and unsteady atmosphere and so its image is much less degraded than that of the object seen at low altitude.

Absorption is self-explanatory: radiation from the celestial body is removed by the air in the line of sight. Scintillation is the effect that causes the stars to appear to twinkle. Turbulent swirls of air of differing temperature, and hence differing refractive properties, move across the line of sight and so slightly deviate the path of the light. As seen in a telescope, scintillation will cause a stellar image to wander rapidly around a mean position and it has the effect of rendering the diffraction pattern invisible. In its place is a rather larger, pulsating image of the star.

Lunar and planetary detail is similarly blurred, and here the image seems to "boil". The effect is rather like the heat shimmer seen along the roads in summer. Unfortunately, scintillation reduces the advantage of large aperture telescopes. From most ordinary locations, even on relatively good nights, scintillation blurs the image so that no detail finer than about 1 arc second can be seen. This is the resolving limit of a good 12 cm telescope! Of course, the lunar and planetary observer is also interested in bright and contrasty images. In order to meet this requirement larger apertures must be used.

For most modern astronomical research the advantage of large aperture telescopes lies in their light grasps. It is because of this requirement that another type of telescope has superseded the refractor: the reflecting telescope.

3.7 The Reflecting Telescope

In the late seventeenth century Sir Isaac Newton began a series of experiments on the nature of light with a view to improving telescopes by the removal of the offending chromatic aberration. In the process he laid the foundations of modern spectroscopy, but he mistakenly considered the false colour effects of the refractor incurable. The solution that he proposed was to replace the lenses in the refractor by mirrors and let these mirrors do the main work of collecting and focusing the light.

The form of *reflecting telescope* that Newton proposed is shown in Fig. 3.8. The light from the celestial body falls down the tube and onto a circular concave mirror (the *primary mirror*) which sends the light back up the tube in a converging cone. Before the rays intersect (and so form a focused image) a flat mirror, inclined at 45°, intercepts the beam and sends it to the side of the tube. In this way the image, viewed with an eyepiece, is brought into an accessible position for the observer. So, in this type of telescope, known as a *Newtonian reflector*, the observer looks in towards the side of the tube, near its top end.

A little before Newton came up with his reflecting telescope the scientist James Gregory put forward his own design. Gregory put a small circular mirror, whose surface was hollowed out into a shape known as an "ellipsoid of revolution", at the top of the tube, instead of the inclined flat mirror of the Newtonian telescope. An ellipse has the property of having two foci. One focus was made to coincide with the focus of the main mirror. The other focus was situated at the bottom of the tube. Hence the elliptical mirror transfers the image from the top of the tube to the bottom of it. (The elliptical mirror of the *Gregorian reflector* and the flat mirror of the Newtonian reflector are both known as *secondary mirrors*.) In the Gregorian telescope the primary mirror is made with a hole in its centre to allow the image to be viewed with an eyepiece.

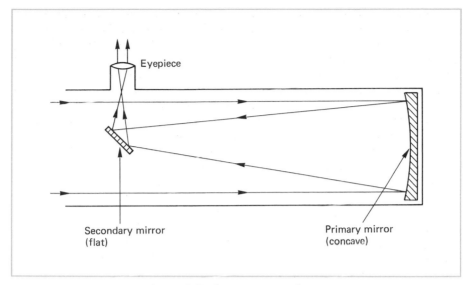

Figure 3.8 The Newtonian reflector.

Two properties of this optical system arose because of its design. One was that the telescope gave an erect image, unlike the inverted images of the retractor and the Newtonian reflector. The other property was that the image formed by the primary mirror was enlarged several times (due to the optical geometry of the arrangement), as well as being transferred to the bottom of the tube. This meant that much larger magnifications would result from the use of a given focal length of eyepiece.

Newton was scathing of Gregory's design when it was first announced to the Royal Society. Newton's reflector was the first to be made (he constructed the first model himself) and it was many years before opticians were able to make the elliptical secondary mirror necessary for the Gregorian telescope.

The Gregorian reflector became very popular in the eighteenth century and the general principle is shown in Fig. 3.9 (*overleaf*), but the telescope is now obsolete. The main reason for this is the development of further designs, notably the *Cassegrain reflector* (Fig. 3.10, *overleaf*). This design is similar to the Gregorian except that the concave, elliptical, secondary mirror is replaced by a convex, hyperboloidal one.

The Cassegrain secondary mirror is placed before the rays from the primary mirror reach a focus, rather than beyond the focus, as is the case for the Gregorian reflector. This results in Cassegrain telescopes having much shorter tubes than is the case for the Gregorians. This is a very important factor when large telescopes are built. In transferring the image to the bottom of the tube the image is enlarged as before but remains inverted. For astronomical use the inverted image is no problem (which way is up in space?) and the extra lenses that could erect the image are left out.

Newton, very unfairly, derided the Cassegrain design as unworkable. It is fair to say that he was rather proprietorial over his reflecting telescope and set about demolishing his rival in a paper for the *Philosophical Transactions* of the

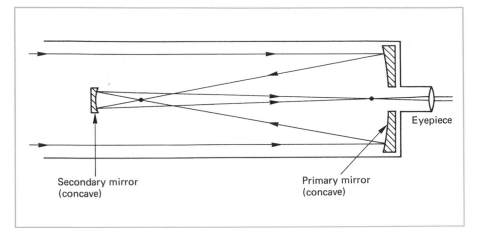

Figure 3.9 The Gregorian reflector.

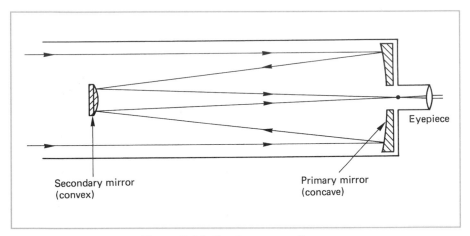

Figure 3.10 The Cassegrain reflector.

Royal Society of May 1672. He concludes his paper with (original spelling and punctuation):

> I could wish, therefore, Mr. Cassegrain had tryed his design before he divulged it. But if, for further satisfaction, he pleased hereafter to try it, I believe the success will inform him, that such projects will be of little moment until they be put in practise.

While Newton's design is still widely used for amateur-sized telescopes (up to 18 inches, or so), for many years now virtually all the largest telescopes at professional observatories have been reflectors of the Cassegrain design. So the Cassegrain reflector has most certainly proved to be of much more than "of little moment"!

3.8 Spherical Aberration

The early telescope primary mirrors, of small aperture and large focal ratio, were given spherical surface figures by the opticians. Such a spherical mirror was adequate to bring all the incident rays to a sufficiently good focus. However, when larger apertures were constructed something better was required. The reason for the shortcoming of the spherical mirror is illustrated in Fig. 3.11. The light rays that fall on the outer parts of a spherical mirror are brought to a focus closer to the mirror than those rays that are incident near the centre.

The result is a blurred image formed at the position of best focus. The image is said to suffer from *spherical aberration*. The remedy is to make the reflective surface of the mirror part of a paraboloid of revolution. Most modern telescopes have paraboloidal primary mirrors. The focusing action of the parabola is illustrated in Fig. 3.12 (*overleaf*)

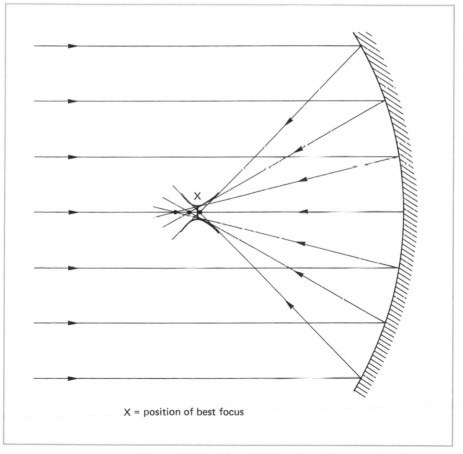

X = position of best focus

Figure 3.11 Spherical aberration.

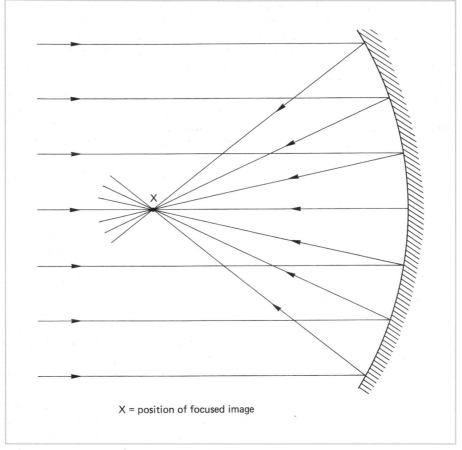

X = position of focused image

Figure 3.12 The focusing action of the parabola.

3.9 The Development of the Reflector

We have already seen how the invention of the achromatic object glass paved the way for large refracting telescopes. Even so, with the 40 inch Yerkes instrument this type of telescope had reached its practical limit of size.

To avoid sagging under their own weight large lenses have to be made thick, but this causes them to absorb a large amount of the light passing through them. Despite their achromatic objectives, large lenses also give a lot of uncorrected colour in the images they form. Large focal ratios (often around f/20) have to be used to reduce this defect to tolerable proportions. This means that refracting telescopes have very long tubes (the Yerkes refractor has a focal length of over 19 metres) and so require massive mountings as well as large domes in which to house them. All this is expensive, and one can have a very much more powerful reflecting telescope for the same outlay.

The early reflectors had mirrors which were made of a material known as speculum metal, an alloy of copper and tin. These metal mirrors were not as reflective as the modern ones and they tended to tarnish rapidly. When first made, these mirrors had to be crafted to extremely high surface accuracies and even a gentle polish would distort the figure. So repolishing meant completely refiguring the optical surface.

Despite this handicap, large speculum metal mirrors were made. In 1789 Sir William Herschel completed his largest telescope, a 48 inch (1.2 m) reflector, which was then the largest in the world. This was surpassed in 1845 by the largest telescope ever to have a speculum metal mirror – the 72 inch (1.8 m) reflector constructed by the third Earl of Rosse, in Ireland.

A silent revolution in instrumental astronomy occurred around 1850 when a method was discovered of chemically depositing a thin silver film onto glass. Not only was the surface so formed much more highly reflective than that of speculum metal but it meant that telescope mirrors could be manufactured from glass. Glass is much easier to grind and polish than speculum metal. Moreover, when the reflective film tarnishes it can simply be dissolved away, using nitric acid, and a new film deposited on the cleaned glass. Nowadays, the silver coating has been superseded by one of aluminium, deposited on the glass by high-vacuum evaporation techniques.

A silver-on-glass reflector of 60 inches (1.5 m) aperture was completed in 1908, followed in 1918 by another of 100 inches (2.5 m) aperture. Both these telescopes are sited at the Mount Wilson Observatory in the USA. The 100 inch reflector remained the largest in the world until 1947, when a truly giant instrument of 200 inches (5 m) aperture was completed at the Hale Observatory on Mount Palomar, also in the USA.

This telescope remained the largest in the world until the 1970s, when the Russians completed their 236 inch (6 m) reflector, though it must be said that this instrument has not been entirely successful. In recent years a new generation of even larger telescopes have been built, but discussion of these will be held over to the next chapter.

Modern large telescopes are designed to be adaptable so that they can be used in a variety of ways. Sometimes the Cassegrain focus is used, but other times the secondary mirror might be removed, equipment then being mounted at the focus of the primary mirror – the *prime focus*. Sometimes the light from the secondary mirror is further manipulated by additional mirrors.

In the *coudé* system an additional flat mirror, known as the tertiary mirror, directs the light through the telescope along its polar axis (this term is explained in the next section). It is normal for the light to be sent *down* the polar axis (which is hollow) but I have used one variant (the 30 inch telescope at Herstmonceux) in which the beam is directed *up* the polar axis to be intercepted by another mirror. The usual form of the coudé arrangement is shown in Fig. 3.13 (*overleaf*). The advantage of this system is that the coudé focus remains fixed, irrespective of how the telescope is pointed. The coudé focus is used where the equipment is too large or to heavy to be attached to the telescope. For some research it is important to house the equipment in a temperature-controlled room.

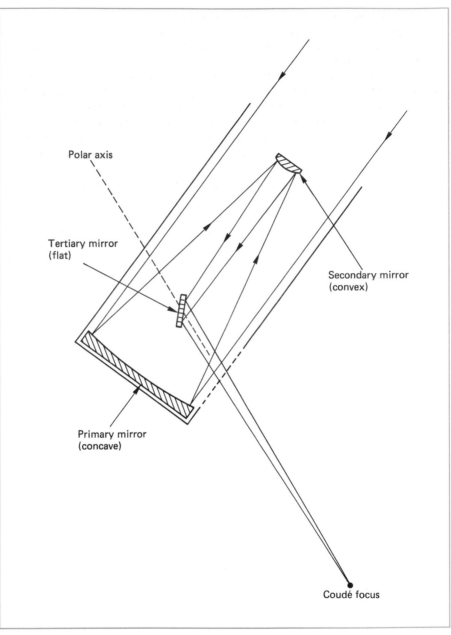

Figure 3.13 The coudé focus.

3.10 Telescope Mountings

If it is to be of much use, even the smallest telescope must have some form of mounting. That mounting should allow two perpendicular motions, in order that the telescope may be pointed to any desired location in the sky. The mounting must be rigid and, for professional and large amateur instruments, it is desirable to have a drive system whereby the telescope can track the celestial bodies in their diurnal motions across the sky.

As stated above, the mounting of a telescope should permit motion about two perpendicular axes. If one of these axes is set parallel to the Earth's rotation axis (in other words, it is aligned to the celestial poles), we then have an *equatorial mounting*. The axis pointing to the celestial pole is termed the *polar axis*. Once the telescope is pointed to its target the other axis, known as the *declination axis*, can be locked and the telescope can track on a celestial body by rotation only about the polar axis.

Many forms of equatorial mounting exist. One type is the *German equatorial*, in which the telescope is counterbalanced by a large weight at the end of the declination axis. The polar axis meets the declination axis at the point of balance (Fig. 3.14). Figure 3.15 (*overleaf*) shows a large refractor on a German equatorial mount.

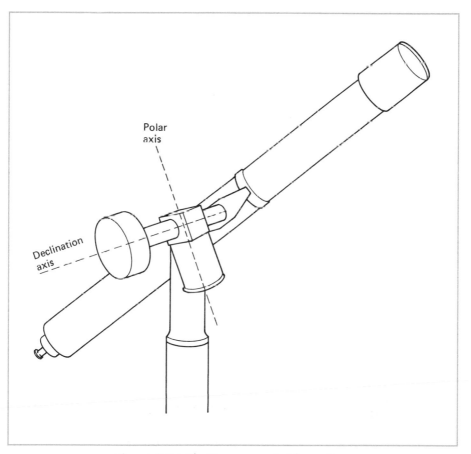

Figure 3.14 The German equatorial mounting.

Figure 3.15 The 26 inch "Thompson refractor" at the Royal Greenwich Observatory, Herstmonceux. This instrument was built by Sir Howard Grubb in 1896. RGO photograph.

Another form of equatorial mounting is the *fork equatorial*. A large cradle, or fork, is attached at the top of the polar axis and the tube of the telescope is slung between the fork arms. The trunnion supports for the tube thus define the declination axis. Figure 3.16 shows my largest telescope, which is mounted in this way. All the other forms of equatorial mounting can be thought of as modifications of the German and fork types.

If one of the axes of a telescope's mount is set vertical, then motions about this axis will cause the telescope to sweep in circles that are parallel to the horizon.

Figure 3.16 The author's 18¼ inch Newtonian reflector. It has a motor-driven fork equatorial mounting. Auxiliary instruments can be seen attached to its skeleton tube.

We say that this is a motion in *azimuth* (see Chapter 1). Motion about the other axis will change the elevation, or *altitude*, of the telescope tube. This arrangement is known as an *altazimuth mounting*. Until a few years ago all large professional telescopes were mounted equatorially. One problem with an equatorial mount is that attaining the desired rigidity is difficult, and hence expensive, due to the polar axis being inclined at an angle equal to the latitude of the observation site. By mounting telescopes altazimuthly this rigidity is achieved at a lower cost. Against this is the disadvantage that simultaneous motions are required about each axis in order to track a celestial body. However, with the advent of computer control this is no longer a problem and almost all new major telescopes have altazimuth mountings.

3.11 The Schmidt Camera

The types of telescope we have been considering so far have been in continual use since their invention. In more recent years other forms of telescope have been

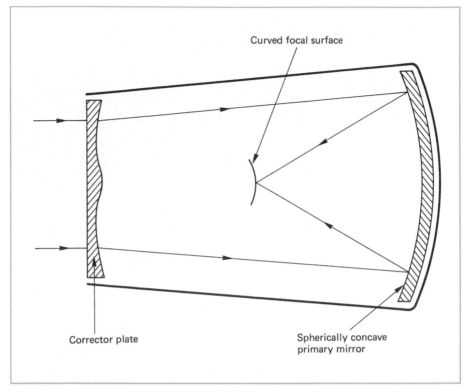

Figure 3.17 The Schmidt camera.

developed which have advantages in their own special fields. The *Schmidt camera* is a case in point. A parabolic telescope mirror gives very good definition near the centre of the field of view but it is incapable of producing sharp images over a large field of view, because of a defect known as coma. In the Schmidt camera a spherical primary mirror is used and the spherical aberration is corrected using a specially shaped lens at the front of the system (Fig. 3.17). Using the Schmidt system astronomers can photograph a very large area of the sky on a single plate. The 48 inch (1.2 m) Schmidt on Mount Palomar takes plates that cover an area of 6° × 6°.

Questions

1   Explain, with the aid of a diagram, the optical principles and construction of a simple refracting telescope. What is *chromatic aberration* and how does it affect the performance of a refractor? How is it remedied? What is the largest refracting telescope in the world? (Give its size and location.) The largest reflecting telescope in the world is much bigger. Give the main reasons why this should be so.

2 What are the main reasons for building large-aperture telescopes? How does the Earth's atmosphere affect the performance of a telescope?

3 Two refractors, one of 40 mm aperture and the other of 160 mm aperture, are used to observe (a) the bright star Vega, (b) a double star consisting of two reasonably bright stars separated by 1.5 arcseconds, and (c) the Moon. Compare and contrast the appearances of these objects in each of the telescopes if both are fitted with 10 mm focal length eyepieces and if both the objectives of these telescopes have focal lengths of 2 m. Calculate the focal ratios of the telescopes and the magnifications in each case.

4 Describe, with the aid of diagrams, the optical principles of the Newtonian and Cassegrain reflecting telescopes. If you were given a 15 cm f/8 Newtonian reflector and had to choose three eyepieces for use with this telescope, explain how you would decide what their focal lengths should be.

5 Compare and contrast *altazimuth* and *equatorial* telescope mountings, emphasising their relative advantages and disadvantages.

6 Calculate the magnifications produced by the following telescopes:
  (a) A 15 cm f/15 refractor, fitted with a 2 cm focal length eyepiece.
  (b) A 30 cm f/5 reflector, fitted with a 1 cm focal length eyepiece.
  (c) An 80 mm f/12 refractor, fitted with a 12 mm focal length eyepiece.
  (d) A 20 cm f/8 reflector, fitted with a 32 mm focal length eyepiece.
  (e) An 80 mm f/20 refractor, fitted with a 4 mm focal length eyepiece.
  (f) A 100 mm f/9.6 reflector, fitted with a 1.2 cm focal length eyepiece.

One of these combinations is unsuitable – which and why?

7 Think about, and write a short essay on, what you consider are the important factors in choosing the site of an optical observatory (e.g. conditions, accessibility, etc.).

8 (a) Explain why large telescopes tend to be reflectors. (b) Two high-quality refracting telescopes, A and B, have object glasses with diameters of 60 mm and 250 mm. Each is fitted with an eyepiece that produces a magnification of ×250. The telescopes are pointed at the same object – two stars that are separated by an angular distance of 1.5 arcseconds. The seeing is near-perfect. Describe the appearances of the object (the two stars close together) as seen through each telescope by an observer with good eyesight (*Hint*: Take special care to emphasise all the **differences** that exist between the two appearances.)

Chapter 4

Modern Developments in Instrumentation

ASTRONOMY may be the oldest science, but astronomers are very modern people. They have been quick to exploit the developing technology in applying new techniques and instrumentation to the study of the Universe.

4.1 The Photographic Plate

Prior to about 1840 the only form of detector available for use with a telescope was the human eye. Then the process of photography was invented. The early plates were very insensitive, allowing the images of only the brightest astronomical bodies, such as the Sun and the Moon, to be registered.

By the end of the nineteenth century photographic plates had improved to such an extent that long exposures could record much fainter objects than could be seen by the observer at the eyepiece of the telescope. For the first time the structures of distant and faint astronomical bodies could be studied.

The science of *spectroscopy* (a method of analysing the light sent to us from the celestial bodies – more fully covered in Chapter 15) arose in parallel with improvements in photography. Eventually photographic plates became sensitive enough to allow the spectra from the celestial bodies to be recorded. Quantitative investigations into the conditions existing in space and the celestial bodies could then be carried out and the science of astronomy progressed rapidly.

4.2 The Photomultiplier

Before the application of photography to astronomy, all measurements of brightness had to be estimates made by eye. The image of a star registered on a photographic plate is a disk (usually much larger than the Airy disk that results from diffraction of the light as it passes through the objective of the telescope) and the size of this disk depends upon the brightness of the star. The brighter the star, the larger the resulting disk image. Measurements of the sizes of these disk images enable stellar brightness to be determined with a much greater accuracy than is possible by eye estimate alone.

A further improvement in the precision of brightness measurements resulted from the discovery of the *photoelectric effect* in 1888. When light falls on the surface of some substances electrons are ejected. If a small disk of this substance is mounted inside a small glass vessel, from which all air has been excluded, then the liberated electrons can be collected by a positively charged electrode (electrons, being negatively charged, are attracted to a positively charged plate) and the resulting electric current can be measured. The brighter the light falling on the photoelectric material, the greater the resulting current.

Many astronomers experimented with the application of the photoelectric effect to the measurement of the brightness of astronomical bodies. A big problem was that the resulting currents were very small because the light levels were so low. A combination of photoelectric detector and amplifier, called a *photomultiplier tube*, was developed in the 1940s.

Figure 4.1 (*overleaf*) shows a simple version of a photomultiplier tube. Light falling on the photoelectric surface causes electrons to be released, as before. These electrons are accelerated towards an electrode which is about 100 volts positive with respect to the photoelectric surface. Each electron striking the first electrode, or *dynode*, causes two or more to leave it. The second dynode is about 100 V positive with respect to the first, causing the ejected electrons to be accelerated towards it. The process repeats itself for each subsequent dynode along the length of the tube. In this way a very feeble light, perhaps causing only one electron to be emitted from the photoelectric surface, generates a virtual avalanche of electrons which reach the final dynode.

A photomultiplier tube coupled to a measuring and recording device is known as a *photoelectric photometer*. These devices can measure the brightness of astronomical bodies to a far greater accuracy than can be achieved using the photographic plate and they are widely used in observatories today.

4.3 Electronic Imaging

Good as the photographic plate is, it does have its limitations. Even the most sensitive plates can register only about one in fifty of the photons of light falling on them. Recent advances in electronic detectors allow fewer of the photons to be wasted. In other words, the telescope can be used more efficiently.

One detector that has proved to be particularly successful is the *charge-coupled device*, or *CCD*. A CCD consists of an array of detecting elements built onto a silicon chip. The values of light intensity falling on each element are read off, usually into a computer, and an image can be reconstructed on a TV screen. Storing the image digitally allows for image manipulation and processing by a computer. In this way features can be emphasised and details brought out that are not apparent in a conventional photograph. Even the most sensitive photographic plates waste 49 out of 50 of the photons falling upon them. In other

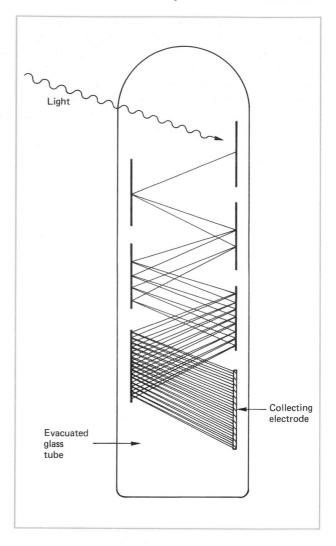

Figure 4.1 A photomultiplier tube.

words, they have a *detector quantum efficiency* (DQE) of 2%. The best CCDs have DQEs of nearly 100% for some wavelength ranges. Clearly, these detectors are utilising the light gathered by telescopes with the maximum possible efficiency.

A disadvantage of CCDs is their comparitively small size when compared with photographic plates. At the time of writing the largest of them measure less than 5 cm square. By contrast, photographic plates of sizes around 25 cm to 40 cm square are in use at some observatories. Obviously the bigger photographic plate will allow a much greater area of the sky to be recorded in a single exposure with a given telescope.

For use at the telescope, many detectors are often cooled to sub-zero temperatures. CCDs and photomultiplier tubes are examples of these. Cooling avoids disruptive thermal effects and allows the detectors to work more efficiently. Even the photographic plate becomes more sensitive at low temperatures.

The *image intensifier* is a development of the photomultiplier tube. Here the image formed by the telescope is focused directly onto the photoelectric surface. The parts of the surface that are most brightly illuminated then emit most of the electrons. The liberated electrons are then accelerated to successive electrodes where the numbers of electrons are multiplied, as in the photomultiplier tube. However, the arrangement is much more complex and the electrons from the final electrode are focused onto a fluorescent screen. In this way an image is formed that is vastly brighter than that collected and focused by the telescope. State-of-the-art image intensifiers today allow a gain of 10^{10} to be reached; in other words, a single detected photon will produce an output of ten thousand million photons

A research team led by Professor Alec Boksenberg developed the *imaging photon counting system*, or *IPCS*, which utilises the best modern detector technology, image intensifiers and a TV camera to feed an image digitally into a computer. The computer then integrates the input to build up an image from the very slow arrival rate of the incoming photons. The DQE of this complete system approaches that of the CCDs.

4.4 Getting More Resolution

At many observing sites atmospheric turbulence blurs an image formed by a telescope, producing a *seeing disk* typically 2 arc seconds across. This means that a star appears expanded into a disk of this angular size. If two stars have angular separations which are smaller than the size of the seeing disk, then they cannot be resolved as two separate objects. They appear as one. In visual wavelengths 2 arc seconds is the resolving power of a telescope of about 7 cm aperture. This means that a 70 cm, or even a 7 m, aperture telescope cannot resolve details finer than can a 7 cm telescope under the same 2 arc second seeing conditions!

It is for this reason that astronomers choose high-altitude sites for their telescopes. Often on the summits of high mountains, the telescopes are then above much of the dirty and unsteady atmosphere. Typical of modern observation sites is Roque de los Muchachos, at a height of 2400 m, on the island of La Palma (one of the Canary Islands). At this site more than 80% of the nights are clear and more than 40% have better than 1 arcsecond seeing. International co-operation has resulted in several major telescopes being set up on the Roque (Fig. 4.2, *overleaf*).

Certain types of measurement can be undertaken even if the measures are of angular distances much smaller than the size of the seeing disk. Early this century the Mount Wilson 100 inch (2.5 m) reflector was fitted with a system of small mirrors at the top end of its tube. The outermost of these mirrors were movable and attached to a long beam. Thus these outer mirrors collected the light from a star and relayed it into the telescope. Ordinarily, the telescope was incapable of resolving the disk of the star, even if the conditions were absolutely perfect. However, the distance between the outer mirrors was such that an interference pattern was produced (this is an effect caused by the wavelike nature of light).

By moving the outer mirrors, the distance separating them when the interference pattern disappears gives a measure of the angular diameter of the star. In this way the diameters of several nearby stars were measured. For example, the

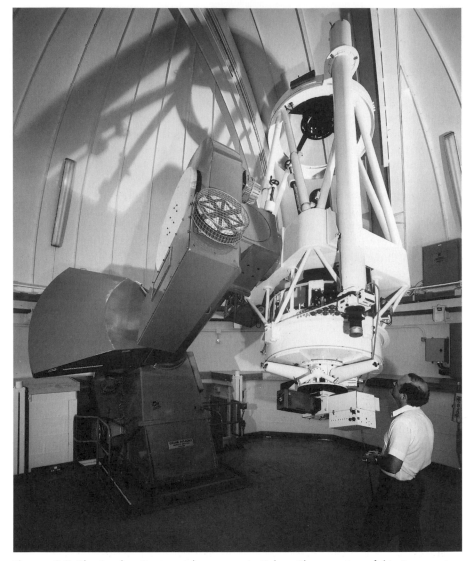

Figure 4.2 The Jacobus Kapteyn telescope on La Palma. The mounting of this 1 m aperture Cassegrain reflector is a modern development of the German equatorial. Note the large counterweight to the lower left. RGO photograph.

star Arcturus was found to have an angular diameter of 0.023 arcseconds. The optical arrangement is known as an *interferometer*.

More recently, Antoine Labeyrie has built an interferometer in France, which consists of two movable telescopes (of 67 m maximum separation). Combining the outputs from these telescopes is not easy, but Labeyrie has succeeded in measuring stellar diameters with this equipment.

Labeyrie is also responsible for another development, called *speckle inter-ferometry*. As we have already seen, the image of a star at the focus of a large telescope is not a tiny point, or even a small disk of a size set by the diffraction pattern. It is a big "blob" because of the effects of atmospheric turbulence. If this "blob" is viewed with a sufficiently high magnification it is seen to be composed of a swirling mass of tiny specks of light. Each of these specks contains information of the finest detail that the telescope can show. Labeyrie has developed a technique of extracting the information contained in a "speckle" and reconstructing an image which shows details as fine as the telescope could show if it were used under perfect conditions.

Astronomers have recently put into operation another method for recovering much of the image resolution lost due to atmospheric turbulence: *adaptive optics*. The method is illustrated in Fig. 4.3 (*overleaf*). A little of the light imaged by the telescope is sampled and the corrugations of the wavefronts of light caused by atmospheric turbulence are detected and analysed millisecond by millisecond by a computer. The computer then drives an array of actuators fixed behind a slightly flexible small mirror. The mirror is made to deform rapidly and in the appropriate way to reduce the wavefront corrugations, and so cancel out most of the image smear that is caused by turbulence. However, this method does have its limitations, one being that a suitably bright star must lie very close in the sky to the object of interest, since the star is needed for monitoring of the required corrections.

4.5 The Latest Generation of Optical Telescopes

For most of the second half of this century the 5 m Hale reflector (the famous 200 inch Mount Palomar telescope) remained the supreme achievement of telescope construction. The performance of the Russian 6 m reflector, completed in the 1970s, was disappointing. Some commentators stated that no larger optical telescopes would ever be built on the Earth (allowing that something bigger might one day be built on the low-gravity environment of the Moon) because of the collossal expense and technical difficulties involved. Indeed, the 1960s, 1970s and 1980s saw a number of telescopes built in the range of 2 m to 4.2 m, but still the Hale reflector reigned supreme in terms of sheer size.

However, even in those later decades, telescope optics manufacturers developed their techniques and were able to produce primary mirrors of low focal ratios, while maintaining, or even improving, their quality. Focal ratios of around f/2 to f/3 were achieved, rather than the ratios of f/3 to f/5 which were common for the larger reflectors of yesteryear.

Note that the largest telescope mirrors were **not** made to be diffraction-limited. The technical difficulties, and hence costs, involved in achieving that accuracy in manufacture would have been far too great. They were generally made accurate enough to be limited mostly by the ambient seeing conditions which they were to be used under. A typical specification for a large telescope mirror was that it should concentrate about 80% of the light from a star into a seeing disk about half an arc second across (some mirrors were a little better than this while some were poorer).

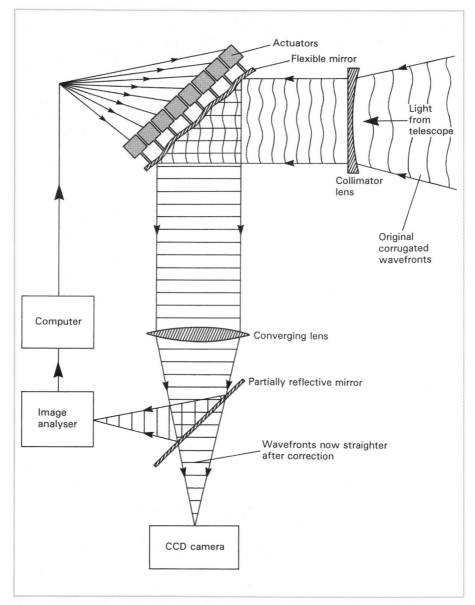

Figure 4.3 Adaptive optics. Highly idealised diagram to show the general principle. For instance the actual magnitudes of the wavefront corrugations and the amount of corrective deformation of the flexible mirror are far too small to show here, as is the spacing of the individual wavefronts.

Engineers also were able to produce more sophisticated designs which saved weight and made for a more compact and much cheaper finished telescope. Computer-driven altazimuth mountings replaced the old equatorials, for instance. This, in turn, allowed large savings in the cost of the building that housed the telescope. Figure 4.4 diagrammatically represents the layout of a modern telescope of about 2.5 m to 4 m aperture.

One recent innovation has compounded the gains in cheapness allowed by the foregoing and has provided another way of improving a telescope's achievable resolution: *active optics*. The mirrors of large telescopes had always to have fairly elaborate mechanical support mechanisms if they were not to distort under their own weight too much. Even so, the mirrors had to be fairly thick (with a thickness of typically one-tenth to one-sixth that of the diameter) and yet the mirrors still tended to distort somewhat and go out of alignment when the telescope was pointed in extreme directions. In active optics, the telescope mirrors are each mounted on a bed of mechanically active supports. A highly sophisticated optical system monitors

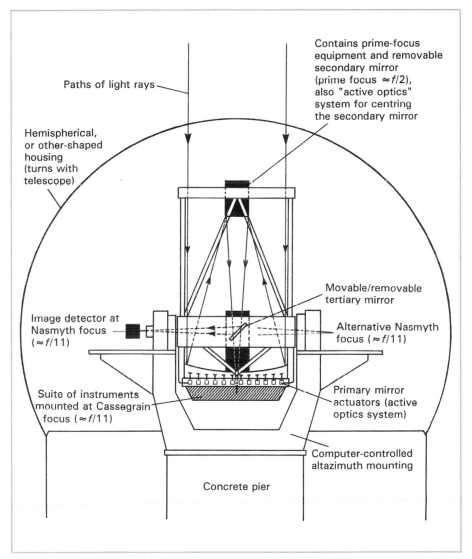

Figure 4.4 The construction of a hypothetical modern telescope of the 2.5 m–4.5 m class, utilising active optics. This diagram only shows some of the main details. Modern telescopes are highly complex electrical–mechanical–optical systems.

the optical performance of the telescope and sends its results to a computer every few minutes. The computer then sends commands to the individual mirror supports to readjust their support pressures as may be necessary. In this way the telescope mirrors are maintained close to their ideal shape and alignment.

In fact, the active optics system is able to correct out much of the inaccuracy remaining in telescope mirrors after their manufacture. In its turn this allows the mirrors to be made very much thinner, and of even lower focal ratio (circa f/1.5 in some cases) and so the whole telescope can be made still lighter! The result of all this is that the latest generation of telescopes are not only of rather better quality, they are also very much cheaper than those of comparable size that were made only a few years ago.

If any one telescope serves as a milestone to represent the advent of the latest developments then it must be the "New Technology Telescope" (NTT) which was completed in 1989. This 3.58 m reflector was built for the European Southern Observatory at La Silla, in Chile. Raw images produced by the telescope often have a resolution of about 0.3 arcsecond, which can be further refined by subsequent image processing by computer. Yet it was built for less than one-third of the cost of the older 3.6 m reflector on the same site while producing images of much higher quality.

All these improvements have opened up the possibility of much larger aperture telescopes. At the time of writing these words, work is under way on several telescopes with apertures of 8 to 8.3 metres. Four of these will form the "Very Large Telescope" (VLT) that is to be completed in the next few years. This will be an array of four identical 8.2 m telescopes set in one long building, in Chile. The light gathered by each can be combined to produce the light-gathering power equivalent to a monolithic 16 m diameter mirror. Also the arrangement will provide the facility of being able to do optical interferometry, and so achieve a very high resolution.

The present record for size is held by the twin 10 m reflectors, known as the "Keck Telescope". Keck I was brought into use in 1990 (although still to be completed at that time) and Keck II joined it on Mauna Kea, Hawaii, in 1996. Each of the 10 m primary mirrors are composed on an array of 36 hexagonal segments. Each segment is mounted on a state-of-the-art support system, with computer-controlled actuators to keep each precisely positioned and aligned. As well as currently being the largest optical telescope system in the world, its 85 m baseline will allow it to be used as a powerful optical (and infrared) interferometer, in the same way as will the VLT.

4.6 The Radio Telescope

It would be a mistake to imagine that astronomers are only interested in the visible part of the spectrum (see Chapter 15 for a full description of the electromagnetic spectrum). Indeed, new knowledge has resulted each time astronomers have opened up another wavelength region. In particular, radio astronomy has rivalled optical astronomy in its contribution to our present state of understanding about the Universe around us.

Radio astronomy was born in 1931 when Karl Jansky, an engineer of the Bell Telephone Laboratories, had been instructed to investigate the sources of radio interference in comunications links. He found that one of these sources originated from the Milky Way. Until then it had not occurred to anybody that radio waves might emanate from celestial bodies. Experiments in radio astronomy began in earnest after the Second World War.

One landmark was the construction of the now famous "Mark 1" radio telescope at Jodrell Bank in Cheshire, England, in 1957. This telescope has a 250 ft (76 m) diameter metal parabolic dish mounted altazimuthly so that it can be pointed to any part of the sky. Incoming radio waves are collected by the dish and focused onto a detector mounted in front of it. The signal is then carried away by waveguides for amplification and processing.

Owing to its fame, this telescope forms the basis of many people's conception of a radio telescope, though many other designs exist. The biggest fully steerable dish in existence today is the 100 m at Bonn in Germany.

At Arecibo, in Puerto Rico, a natural crater in the landscape has been used to create a 305 m fixed dish type of radio telescope. The crater was originally covered in a fine metal mesh, later replaced with aluminium panels, and a detector suspended high above its surface. The rotation of the Earth allows a narrow strip of the sky to be scanned. This is the largest single-dish type of radio telescope in the world. It has proved highly successful and has provided astronomers with many discoveries and important results and at the time of writing these words it is being further upgraded.

Another form of radio telescope consists of an array of individually steerable aerials, each gathering the extremely weak signal that arrives from the astronomical source.

The resolution of a radio dish is much poorer than that of an optical telescope due to the much longer wavelengths of radio waves. (To see how wavelength affects resolution inspect the formula for the Rayleigh limit given in the last chapter.) Interferometry can be applied to radio astronomy in the same way that it can to optical astronomy. The outputs of two or more radio telescopes are combined to synthesise the resolution of a single dish of diameter equivalent to their separation.

Nowadays, radio interferometers are very widely used in astronomy. Even the Mark 1 radio dish at Jodrell Bank (now named the Lovell Telescope, after Sir Bernard Lovell, the British pioneer of radio astronomy and head of the group that created the instrument) has been linked with several other radio telescopes spread across the English countryside to form MERLIN (for Multi Element Radio Linked Interferometer Network). It is even possible to link telescopes in different countries and so synthesise the equivalent of a radio dish the size of the Earth! This is called *VLBI* (for Very Long Baseline Interferometry). A single dish may well have a radio resolution inferior to the human eye working at visual wavelengths but using VLBI radio astronomers are able to achieve resolutions greatly superior to the best modern optical telescopes!

A radio telescope does not produce a direct image as does an optical telescope. Instead, it produces an amplified signal of the radio emission from a patch of the sky. The patch has a size that is determined by the resolution of the telescope for the wavelength at which it is working. Depending on the type of observation, a record of the variations of the strength of the signal with time might be produced as a trace on a pen recorder, or the telescope might be set to track over a small area of the sky around the object. A *radio map* of that area might then be produced. This is a "picture" of the area in radio emission and is often artificially coloured to indicate different levels of radio intensity.

4.7 Astronomy at Other Wavelengths

Most radio telescopes operate at wavelengths not much shorter than a metre. However, some instruments have been designed to work in the microwave region of the spectrum, where the wavelengths are in the centimetre range. These give information on the chemical compositions of the large clouds of matter that permeate space.

Going to shorter wavelengths still, many telescopes have been designed to work in the infrared part of the spectrum. Infrared radiation is emitted from all warm or hot bodies and this causes a problem – special detectors, often cooled to near absolute zero (–273 °C) are needed. Also, the Earth's atmosphere, as well as the telescope and its surroundings, emit very strongly in just this frequency range. Observing in the infrared has been compared to trying to carry out an optical investigation of a faint object in full daylight and with a luminous telescope!

Special techniques have to be used to separate the faint signal from the astronomical source from the much greater background radiation. Nevertheless, infrared astronomy has added a vast amount to our knowledge, particularly of those objects that are at relatively low temperatures and so are much less luminous, if optically visible at all.

4.8 Astronomy from Above the Ground

The Earth's atmosphere, while being necessary for human survival, is a distinct nuisance to the work of astronomers! As well as the effects on optical work, the atmosphere is largely opaque to much of the radiation arriving from space. In some instances going to very high mountain sites, where the air is thin and clear and relatively devoid of water vapour, does help. Thus an infrared observatory has been set up atop an extinct volcano on Mauna Kea, in Hawaii, at 4200 m above sea level – so high that many astronomers who use it suffer from altitude sickness.

Balloons and aeroplanes have been, and are, used to carry telescopes aloft but really to open up the whole of the spectrum for study it is necessary to get right above the Earth's atmosphere and out into space. To date, a great many satellites and spacecraft have carried astronomical equipment and many missions have been devoted solely to astronomical research. One notable achievement was *IRAS* (InfraRed Astronomical Satellite), which operated in orbit about the Earth and during 1983 surveyed the entire sky at infrared wavelengths and even discovered several new comets. The space-borne infrared studies of IRAS have been followed up by the Infrared Space Observatory (*ISO*), launched in November 1995.

Ultraviolet, X-ray and gamma-ray studies have also been undertaken from above our atmosphere and some satellites and space probes have been given specific targets, such as Earth sciences and weather monitoring (for instance Meteosat) and the Sun (the Solar and Heliospheric Observatory, or *SOHO*).

Of course, the crowning achievement of space-borne astronomy has to be the Hubble Space Telescope (HST). Orbiting above the Earth's atmosphere, its 2.4 m mirror feeds several different types of instruments and it has already produced a large number of new discoveries about the Universe, as well as increasing our general knowledge considerably.

However, its ride has been far from smooth. For a variety of reasons its launch was delayed several years and when it was finally taken aloft by the Space Shuttle *Discovery* in April 1990 it didn't take the ground crews long to find that the optical system had a major flaw. Investigations unearthed that the manufacturers of the hugely expensive primary mirror had made a mistake in relying on a piece of test equipment that itself had turned out to be faulty. It had been claimed that the HST mirror was the most accurate of that size ever made – and so it was. The trouble was that it was accurately the wrong shape! Against the background of the world's Press and the 1.5 billion dollar price tag of the HST, faces went red and I imagine that heads rolled. Observations could begin, but a repair mission was needed to fit the HST with corrective optics, as well as attending to a number of mechanical problems. Now the HST performs brilliantly, routinely delivering images of 0.1 arc second resolution in visible wavelength; it can also observe and measure radiations that are blocked by our atmosphere.

From this chapter it may be apparent that one of the technological growth areas that has particularly benefited astronomy is that of computers. These are now used extensively for instrument control, as well as for data aquisition and processing. Indeed, the advent of powerful computers has led to a new revolution in astronomy, which already promises to be every bit as important as that created by the invention and subsequent development of the telescope.

This review of modern techniques and instrumentation has necessarily been sketchy. The discoveries alluded to are dealt with in their due places in later chapters of this book but, for now we can see that we have come far since Galileo's tiny "optick-tube" of nearly four centuries ago.

Questions

1. Outline the various types of detector used with telescopes operating in the visual part of the spectrum. Compare and contrast their usefulness, paying particular attention to their relative sensitivities.

2. Explain how a photomultiplier works and outline its use in astronomy.

3. Write an essay on radio astronomy.

4   What is *interferometry* and how has it been used in optical and radio astronomy?

5   Write a short essay on infrared astronomy.

6   How has the use of balloons, aircraft and satellites been of benefit to astronomy?

7   Find out about, and write a short essay on, the use of computers in astronomy.

8   Outline the main reasons why the Hubble Space Telescope is so important to the work of astronomers.

9   Outline the problems faced by astronomers working in the following areas: radio astronomy; infrared astronomy, optical astronomy, ultraviolet astronomy, low-orbit satellite observing (all wavelengths). In cases where there are remedies for any of these problems, explain what they are.

10 Outline the main improvements in optical telescope technology since the 1970s.

Chapter 5

Gravitation

THE Universe is composed of a mixture of matter and energy. As far as we know at present, just four basic types of force give our Universe its structure and properties. Three of the four – the strong nuclear, the weak nuclear and the electromagnetic/electrostatic – are mainly concerned with the small-scale structure of matter. These forces are strong but operate over very small ranges. The other force is gravitation. In terms of magnitude, gravitation is by far the weakest of these natural forces and yet, because of its range of influence, it has shaped our Universe, on the large scale, into its present form.

5.1 Kepler's Laws of Planetary Motion

The work of Copernicus and Galileo established that the planets of our Solar System orbit the Sun. However, a few details needed to be cleared up. The Copernican view was that the orbits of the planets were circular in form. As observations of the positions of the planets became more precise it was clear that this could not be entirely true. Notable among the early positional astronomers was the Dane Tycho Brahe, working in the late sixteenth century. Using elaborate quadrants and sextants Tycho was able to measure the positions of the planets and stars to accuracies of one or two minutes of arc.

The planets were found not to stick to their predicted paths. The puzzle was only solved in 1609 when the German mathematician Johannes Kepler, after a careful analysis of the observations of Tycho Brahe, realised that the paths of the planets about the Sun are actually ellipses. In 1609 Kepler published two "laws" of planetary motion, followed by a third in 1618:

Law 1

The planets move in elliptical orbits, with the Sun located at one of the foci.

Law 2

The radius vector from the Sun to the planet sweeps out equal areas in equal intervals of time.

Law 3

The squares of the sidereal periods of the planets are proportional to the cubes of their mean distances from the Sun.

Law 1 is fairly easy to explain. One way of drawing an ellipse is to mount a piece of paper on a board, stick two pins in the board, put a loop of thread over the two pins and pull the thread tight with a pencil. The pencil can then be moved keeping the string taut. The shape drawn out is an ellipse. The pins form what we call the *foci* of the ellipse. If the foci are close together the shape formed will be nearly a circle. If the foci are far apart a long, narrow, ellipse will be formed. In the case of the planets, the Sun is at one of the foci. The other is in empty space.

Law 2 is easy to understand if reference is made to Fig. 5.1. If the orbit is elliptical then it follows that the distance from the planet to the Sun will vary as the planet goes about its orbit. When the planet is close to the Sun it will be travelling faster than when it is moving further out. When the planet is closest to the Sun we say that the planet is at *perihelion*. When at its furthest, we say the planet is at *aphelion*. We call the line joining the centre of the planet to the centre of the Sun the *radius vector* of the planet. In Fig. 5.1 the planet moves through a distance AB in a certain time whilst it is near perihelion, and thus sweeps out an area ABS in that time.

Similarly, the planet moves through a distance CD in the same interval of time. In that time an area CDS is swept out. It turns out that these two areas are equal. Thus the radius vector sweeps out equal areas in equal times.

Law 3 states that there is a definite relationship between the sidereal period of a planet (the time that it takes to go once round the Sun) and its mean distance

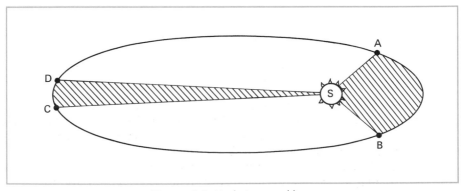

Figure 5.1 Kepler's second law.

from the Sun. (Actually this is the form of the law generally given at the elementary level. In truth the relationship is between the sidereal period, as stated, and the *semi-major axis*. However, to explain that would involve a discourse in the geometry of conic sections. Hence I have confined myself to including the usual elementary definition, which is adequate for most purposes. In this context the mean distance is calculated by adding the perihelic and aphelic distances together and dividing by 2.) Mathematically expressed the relationship is:

$$d^3 \propto T^2$$

where d is the mean distance of the planet from the Sun and T is the sidereal period of the planet. If both T and d are known for one planet, then we may calculate T or d for any other, knowing either quantity. For instance, if we know the mean orbital radius of the Earth and its sidereal period, we could find the mean orbital radius of another planet if we knew its sidereal period.

5.2 Newton's Law of Gravitation

It is said that while Newton was sitting in his garden one day he chanced to see an apple fall from a tree. On reflection, he realised that the same force that pulled the apple to the ground was responsible for keeping the Moon in orbit about the Earth. He thought it likely that the same sort of force operated between the planets and the Sun, keeping the planets in their orbits. Whether this story is true or not, Newton developed the necessary mathematics to expand his theory and in 1687 he published the result in his masterly work, the *Principia*. Newton was able to formulate a law which he believed would hold true anywhere in the observable Universe. *Newton's universal law of gravitation* states:

Any two bodies will attract each other with a force which is proportional to the product of the masses and is inversely proportional to the squares of the distances separating them.

The law can be expressed in equation form:

$$F \propto Mm/r^2$$

or $$F = GMm/r^2$$

where F = the attractive force, measured in newtons,
  M, m = the masses of the bodies, measured in kilograms,
  r = the distance of separation, measured in metres,
  G = a constant of proportionality, known as either the *universal constant of gravitation*, or the *gravitational constant*.

From experiments the value of G was found to be 6.67×10^{-11} N m^2 kg^{-2}.

If at least one of these masses is large, then the force of attraction will be large. The mass of the Earth is 6×10^{24} kg and is thus responsible for the large forces exerted even on small masses on its surface.

Let us use the above equation to predict the gravitational force exerted on a 1 kg mass on the surface of the Earth. Let M be the mass of the Earth and m be our 1 kg mass. r is the radius of the Earth (we are considering a body on the Earth

– this is the same as saying that it is one radius away from the Earth's centre). Using the value of G and M given above, F comes out to be 9.8 N (9.8 newtons). In other words, a body of mass 1 kg will be subject to a gravitational force of 9.8 N on the Earth's surface. It will weigh 9.8 N. A mass of 2 kg will be subject to twice the attractive force. This body will weigh 19.6 N – and so on.

5.3 Circular Motion

Before dealing with escape velocity and satellite orbits we must consider a little of the mathematics of bodies that are moving in circular paths while under a central force. Much of the following also applies to bodies moving in elliptical paths, but we will take the mathematically much simpler case of circular motion.

One unit of measurement useful to us is the *radian*, so we will begin by defining this. Imagine a circle whose circumference is divided up into a number of arc lengths. Now, the radius of a circle is equal to 2π times its radius ($\pi = 3.142$). If each of these arc lengths is equal to the radius of the circle then there must be 2π (6.284) arc lengths fitted around the circumference of the circle. The angle, measured from the centre of the circle, which one of these arc lengths subtends is known as 1 radian (Fig. 5.2). It should be obvious that there are 2π radians in 360°.

Consider a body, of mass m, moving in a circular path of radius r (Fig. 5.3, *overleaf*). If there is nothing to stop it, we may suppose that the body will continue in its motion. However, a force F must be present. That force has to be directed from the body towards the centre of the circle in order to keep the body in its circular path. No other single and unvarying force will keep the body moving in a circle. Take away that force and the body will move off in a direction at a tangent to the circle. (Imagine what happens if you are twirling a stone, attached to a piece of string, in a horizontal circle above your head and then you let go.)

Let us suppose that the body moves once around the circle in a time T seconds. T is thus the *period* of the body about the centre of the circle.

The angle, as measured from the centre of the circle, through which the body moves per unit of time (for our purpose per second) we call the *angular velocity*, ω, of that body. Since there are 2π radians in a complete circle, we can say that:

$$\omega = 2\pi/T$$

At any instant the body will have a velocity, v, tangential to the circle given by:

$$v = r\omega$$

If the force, F, suddenly vanishes then the body will move off, at a tangent to the circle, with this velocity.

As the body moves around the circle, the direction of its tangential velocity is always changing. This comes to the same as saying that the body is accelerating towards the centre of the circle (even though the radius of the path is constant, as is the speed of the body).

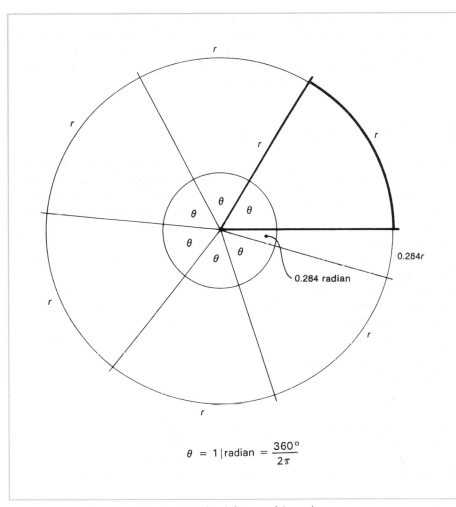

$$\theta = 1\,\lvert\text{radian} = \frac{360°}{2\pi}$$

Figure 5.2 The definition of the radian.

Using vector analysis (beyond the scope of this book) it can be shown that:

$$a = r\omega^2$$

where a is the resultant acceleration.

Newton's second law of motion states that $F = ma$, where F is the force acting on a body, of mass m, and a is the resultant acceleration. Moreover, the acceleration takes place in the same direction in which the force acts. In the case of the body experiencing circular motion both the force and the acceleration are directed towards the centre of the circle. Thus:

$$F = mr\omega^2$$

The force, F, is known as the *centripetal force* and, in the case of a planet orbiting the Sun, or a satellite orbiting a planet, it is provided by gravitational attraction.

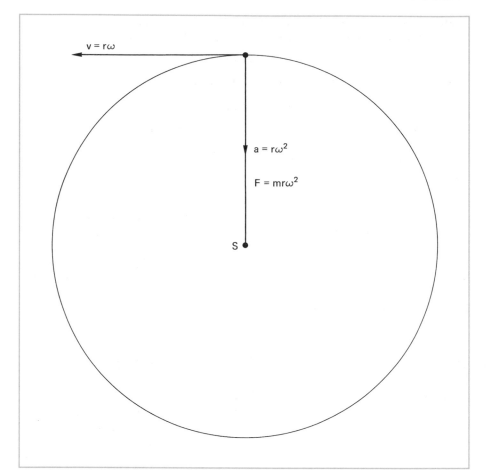

Figure 5.3 A body undergoing circular motion. The velocity of the moving body is always tangential to the circle but the centripetal force and acceleration are both directed towards the circle's centre.

Kepler used Tycho Brahe's observations to empirically derive his third law of planetary motion. We can use the above result to predict it mathematically.

5.4 Mathematical Derivation of Kepler's Third Law

Consider a planet, of mass m, orbiting the Sun, of mass M, at a distance r. Let the sidereal period of the planet be T.

From Newton's universal law of gravitation we have:

$$F = GMm/r^2$$

where F is the gravitational force of attraction between the Sun and the planet. If we assume that the Sun is much more massive than the planet (which is the case for all the planets in our Solar System), so that we can consider the Sun to be stationary while all the movement can be ascribed to the planet, then we have:

$$F = mr\omega^2$$

where ω is the angular velocity of the planet in its orbit.

Thus, $$GMm/r^2 = mr\omega^2$$

But $\omega = 2\pi/T$, so substiting, cancelling and rearranging gives:

$$r^3/GM = T^2/4\pi^2$$

Hence, $$r^3 \propto T^2$$

This relation can be compared with the statement of Kepler's third law, already given. Notice that the mass of the planet does not affect its orbital period. This is only true where the mass of the planet can be taken as negligible compared with that of the parent body (in our case the Sun). If this is not so then the two bodies must be considered to rotate about their common centre of gravity. It is true that we have used the theory of circular motion to derive Kepler's third law, when the true orbital paths are ellipses. However, a rigorous mathematical derivation shows that the end result is the same.

5.5 Two-Body Orbiting Systems

The Earth and Moon form a typical system of two orbiting bodies. It is commonly said that the Moon orbits the Earth, but this statement needs qualification. It is true to say that both bodies orbit about their common centre of mass. The Earth is 81 times as massive as the Moon. This means that the common centre of gravity, or *barycentre*, lies $\frac{1}{82}$ of the way along the line joining the centre of the Earth to the centre of the Moon.(The ratio of distances to the barycentre is 1 : 81.) Thus the barycentre lies inside the globe of the Earth, so the original statement is true but only to a first approximation.

Not all two-body orbiting systems will have one component that is much more massive than the other. Indeed, a great many stars are known to be gravitationally bound into mutual orbits and these *binary stars* are usually much more even. The barycentre is then in mid-space, but shifted towards whichever star is the more massive. The distances from the centre of each star to the barycentre are in the inverse ratio of the masses. Figure 5.4 (*overleaf*) should make this clear.

5.6 The Tides

It was Isaac Newton who explained the twice daily tides in terms of the gravitational interaction of the Moon with the Earth. The force of attraction of the Moon on the oceans of the Earth causes a bulging of the oceans in the direction of the Moon. In effect, the water is being "heaped up" by the attraction of the Moon. The Earth itself is also being pulled away from the water on its reverse side. The result is that another bulge of water forms on the side of the Earth opposite to the Moon.

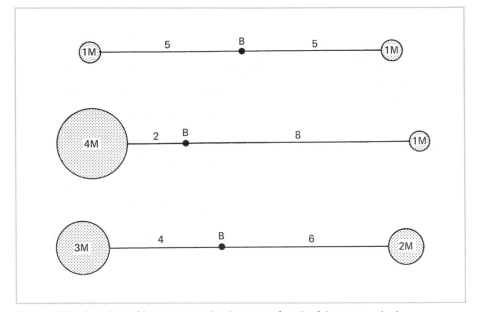

Figure 5.4 The orbits of binary stars. The distances of each of the stars to the barycentre are in the inverse ratios of their masses in each of these cases.

As the Earth turns on its axis (one could say that the Earth "turns under the Moon") so each position on the Earth experiences two tides per day when the level of the sea along the coastline appears to rise. The principle is illustrated in Fig. 5.5.

The Sun also exerts a tidal influence on the oceans of the Earth but, because of the much greater distance of the Sun compared with the Moon, the Sun produces only about half the effect on the Earth's waters that the Moon does. When the Moon and the Sun both pull along the same straight line (either in the same direction, or in opposite directions) the Earth experiences tides of greatest amplitude. These are known as *spring tides* (Fig. 5.6). As will be seen in the next chapter, this occurs at the time of New Moon and Full Moon.

When the Moon and the Sun pull in perpendicular directions, the Earth experiences its weakest tides, known as *neap tides* (see Fig. 5.7, *overleaf*). Neap tides occur around the times of first and last quarter Moon. It must be borne in mind that the picture of tides given here is, necessarily, a simplified one and such effects as sea currents and coriolis forces (due to the spin of the Earth), as well as local topography, must be taken into account in a full description of tidal effects.

5.7 The Measurement of the Astronomical Unit

In the past considerable efforts have been made to determine accurately the distance of the Earth from the Sun. Once this distance is known the true distances to the other planets can be found. Also, the Earth–Sun distance forms the baseline

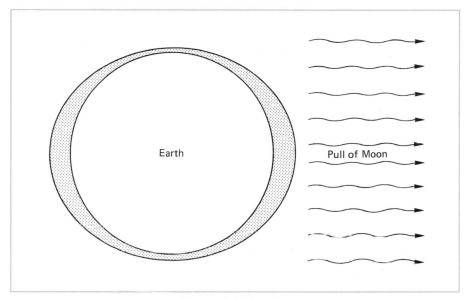

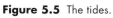

Figure 5.5 The tides.

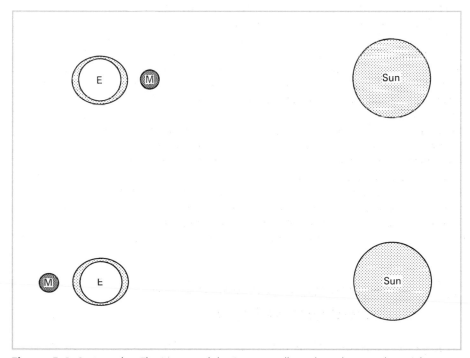

Figure 5.6 Spring tides. The Moon and the Sun are pulling along the same line and so give rise to tides of maximum amplitude.

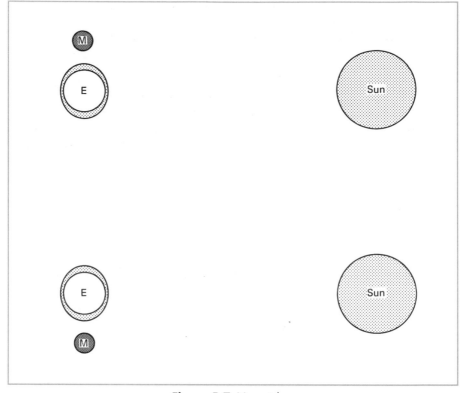

Figure 5.7 Neap tides.

for a method of finding the distances to the stars. Other information, such as the mass of the Sun, can also be determined.

The principles of radar were applied to radio astronomy in the 1960s. If a sharp radio pulse is broadcast from a radio telescope pointed in the direction of a solid object and the time taken for its faint echo to arrive is measured, then, knowing the speed of the radiowaves, the distance of the object can be calculated.

The Sun is not a solid body but the planet Venus is. The radar technique was used to find the Earth–Venus distance. The sidereal periods of the Earth and Venus were accurately known. The application of Kepler's third law gave the ratio of the Earth–Sun and Venus–Sun distances and so the Earth–Sun distance could be found. Owing to its importance, the value of the Earth–Sun distance, 150 million km, is known as the *astronomical unit (AU)*.

5.8 The Earth's Gravitational Field

It is common experience that a stone, if thrown upwards, will rise to a certain height and then fall back to the ground. Hence the saying "What goes up must come down." However, if the stone is fired up at a velocity of 11.2 km/s then this statement is no longer true. Neglecting the effects of friction with the air, the stone will leave the Earth never to return. This velocity is commonly called the *escape velocity* of the Earth.

The energy which a body has when it is in motion is called *kinetic energy* and is given by the formula:

$$\text{Kinetic energy} = \tfrac{1}{2}mv^2$$

where m is the mass of the body, measured in kilograms, and v is its velocity, measured in m/s. The energy is then measured in joules.

A body in the Earth's gravitational field, on the surface of the Earth, is endowed with a type of stored energy, known as *gravitational potential energy*, which is given by the following equation:

$$\text{Potential energy} = -GMm/r$$

where G = the universal constant of gravitation,
$\quad M$ = the mass of the Earth,
$\quad m$ is the mass of the body,
$\quad r$ = the radius of the Earth.

The minus sign tells us that energy has to be given to the body in order to increase its distance from the centre of the Earth. We will leave it out and only consider the modulus of the gravitational potential energy for the following argument.

In order for an object to leave the Earth's gravitational field, we must give it enough kinetic energy to equal, or exceed, its gravitational potential energy. Thus:

$$\tfrac{1}{2}mv^2 = GMm/r$$

Cancelling, $\qquad\qquad \tfrac{1}{2}v^2 = GM/r$

so $\qquad\qquad\qquad v^2 = 2GM/r$

$$\therefore \quad v = \sqrt{2GM/r}$$

It should be apparent from the above equation that the escape velocity of a planet is solely dependent on its mass and radius. For the Moon the escape velocity is 2.4 km/s.

The average velocity of the molecules of a gas at normal pressures and temperatures is around 0.5 km/s. Some of the molecules move more slowly than this average figure and some move much faster. There is a given statistical distribution of the speeds of molecules that is independent of the amount of gas considered. This is the reason why the Moon has not been able to retain any appreciable atmosphere, while the Earth, with its greater escape velocity, has. Certainly, though, the Earth has lost nearly all the hydrogen and helium that we believe dominated its early atmosphere, as these gases possess light molecules and so move with high velocities.

5.9 The Acceleration Due to Gravity

Legend has it that Galileo one day performed an experiment by dropping two metal spheres, one small and one large, from the top of the famous Leaning

Tower of Pisa. Both spheres were released at the same instant and both hit the ground at the same time. This showed that, neglecting the effects of air resistance, bodies undergo the same rate of acceleration when they fall freely near the surface of the Earth, no matter how their masses differ.

This acceleration is known as the *acceleration due to gravity* and has a value of 9.8 m/s² (metres per second squared) near the surface of the Earth. The acceleration due to gravity is often given the symbol g. The value of g on the Moon's surface is 1.6 m/s². That the acceleration due to gravity on the surface of a planet depends solely on the mass and radius of that planet can be shown using Newton's second law of motion and his universal law of gravitation.

Newton's second law of motion is usually expressed as the equation:

$$\text{Force} = \text{Mass} \times \text{Acceleration}$$

In the case of the object falling freely under gravity, the accelerating force is the planet's gravitational attraction on the body and the acceleration that arises as a consequence is g. If we let the mass of the body be m and the accelerating force be F, then we can say:

$$F = mg$$

We can use the universal law of gravitation to give this accelerating force in terms of the mass of the planet, M, its radius, r, and the universal constant of gravitation, G:

$$F = GMm/r^2$$

Thus, $$mg = GMm/r^2.$$

Cancelling, $$g = GM/r^2$$

We have already seen that the escape velocity, v, is given by:

$$v = \sqrt{2GM/r}$$

So we can put the escape velocity into terms of the acceleration due to gravity:

$$\therefore \quad v = \sqrt{2gr}$$

When a body is not in free fall but is at rest, it will exert a force F, given by $F = mg$, on whatever supports it. This force we know as the weight of the body.

5.10 Rockets and Satellites

If a certain date marks the beginning of the "space age", then that date must be 4 October 1957. On that day the Russian satellite Sputnik 1 was sent into space to become the first new body in orbit about the Earth in 4600 million years! Nowadays we take communications and weather satellites for granted. Although it took mankind three centuries to develop the technology, it was Isaac Newton who originated the initial concept of placing a body in orbit about the Earth. Figure 5.8 is an illustration of his basic idea.

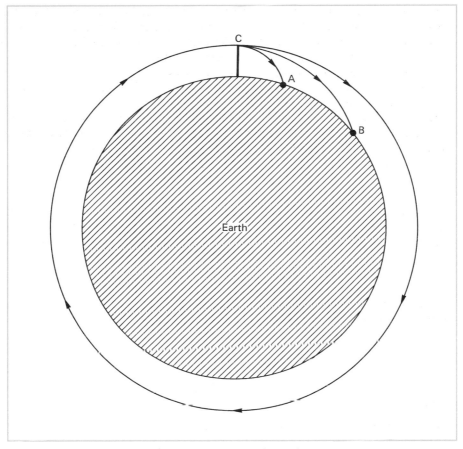

Figure 5.8 A projectile in orbit.

Referring to the diagram, if a body is thrown off a tower it will travel with a parabolic path hitting the ground at *A*. If the body is given a greater horizontal velocity, then it will travel a greater horizontal distance, hitting the ground at *B*. If we neglect any aerodynamic effects, then the vertical accelerations of the bodies are the same in each case. If the body is given a great enough horizontal velocity it would be found that the surface of the Earth itself "falls away" at the same rate that the body falls towards the centre of the Earth.

In other words, the object would be continually accelerating towards the ground but would never actually reach it! This is another example of how a body in circular motion accelerates towards the centre of the circle. The body is said to be in *free fall*.

If we consider a passenger-carrying lift where the cables break and the lift cage plunges down the lift shaft in free fall, the unfortunate passengers become "weightless" until the cage hits the bottom of the shaft. That is, they exert no weight on the floor of the cage whilst it is descending. The occupants of a space-craft in orbit are just as much in free fall and so they experience the same sensation of weightlessness. Weightlessness has nothing whatever to do with "being out of the Earth's gravity", or "being above the atmosphere", or any one of a dozen other common misconceptions.

Needless to say, a satellite is not fired from a tall tower by some sort of power-ful cannon. It is put into orbit by the use of a rocket which is initially set in a vertical direction but then follows a carefully computed path until, when it is above the atmosphere and is at the correct height and speed, it is travelling parallel to the ground. The satellite is then released.

The rocket operates on the "reaction principle". It injects its fuel (usually liquid hydrogen) into a chamber where it is mixed with oxygen (also carried on board the rocket) and is ignited. The resulting hot gases are exhausted at extremely high velocities. Thus the rocket exhaust has a large momentum in the "backwards" direction. This is balanced by the rocket gaining momentum in the "forwards" direction.

Unfortunately, no large, single-stage, rocket laden with fuel would ever be able to get into space because of its own weight and so the principle of the "step rocket" was developed. The rocket is divided up into separate stages (exactly how many depends upon the rocket and its intended payload) and these stages are fired sequentially, the next being fired when the fuel in the previous stage has been used up and that stage has been jettisoned. In this way the latter stages of the rocket have a much lighter load to accelerate up to orbital velocity. Figure 5.9 is a very simplified illustration (using a fair amount of artistic licence!) of the general principles of launching a rocket and placing its payload in orbit.

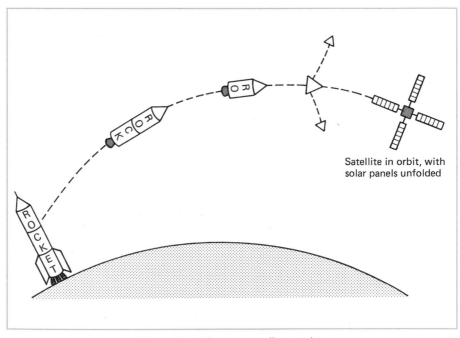

Satellite in orbit, with solar panels unfolded

Figure 5.9 Placing a satellite in orbit.

Questions

1   State *Kepler's laws of planetary motion.* (Illustrate your answer with diagrams where necessary.) Define the terms *perihelion* and *aphelion.* The planet Earth orbits the Sun at a mean distance of 150 million kilometres and has a sidereal period of 1 year. The planet Jupiter has a sidereal period of 11.9 years. What is the mean distance of Jupiter from the Sun, (a) in kilometres and (b) in astronomical units?

2   State *Newton's universal law of gravitation.* Express the law in terms of an equation, defining all the terms used. Show how the law can prove Kepler's third law of planetary motion.

3   Describe **fully** the motions of the Earth and Moon with respect to one another and state what differences would arise if the Moon and Earth were equal in mass.

4   Explain how the tides in the Earth's oceans are caused. Explain the difference between *spring tides* and *neap tides.*

5   Explain what is meant by *escape velocity* and derive equations for the escape velocity in terms of (a) the universal constant of gravitation and (b) the acceleration due to gravity. Under what conditions would a feather fall to the ground at the same rate as a hammer?

6   Briefly explain how a satellite is put into orbit about the Earth. Some satellites that have highly elliptical orbits (or those with low orbits) eventually burn up in the Earth's atmosphere. Explain, as fully as you can, why you think this happens.

7   Satellite A is in a "stationary orbit" above the Earth (think carefully what this statement actually means!). What is the orbital period of satellite B, if B's orbit is twice as far from the centre of the Earth as A's orbit?

8   You notice a point of light moving steadily, and silently, across the night sky. Apart from its motion it looks very much like one of the brighter stars also on show. It takes only a minute or so from you noticing it in the south-west to it vanishing while still quite a way above the horizon in the north-east. (a) What is the object likely to be? (b) Why do you think it vanishes from view before it reaches the horizon? (*Hint:* Think about things casting shadows.)

Chapter 6

The Moon

THE Moon is the Earth's only natural satellite. To the naked eye it is one of the most splendid of the heavenly bodies and it has held the fascination of mankind for thousands of years. It has long been known that the tides were related to the position of the Moon, and the highwaymen of old certainly concentrated their activities around the time of full Moon, when the illumination that it provided was at its greatest. Various gods have been associated with the Moon since ancient times. Many superstitions are connected with the Moon. It is supposed to be unlucky to look at the crescent Moon through glass (some people say "the new Moon", but the new Moon cannot be seen at all – they really mean the thin crescent Moon!). It is commonly held that insanity increases at the time of the full Moon. Probably connected with this is the legend of lycanthropy!

However, scientists tend to be rational people and we have learned much about our cosmic neighbour, especially since Man first set foot on its surface, more than a quarter of a century ago.

6.1 The Nature and Origin of the Moon

Despite the fact that the Moon is often a brilliant object in the night sky, it does not actually emit any light of its own. It shines purely by reflecting light from the Sun. The Moon is a hard, rocky body with an equatorial diameter of 3476 km and, as such, is over a quarter of the diameter of the Earth. This has led many astronomers to consider the Earth and Moon as a double planet system, rather than as a parent body–satellite system. Certainly all of the other planetary satellites in the Solar System, apart from that of the peculiar Pluto, have diameters which are a much smaller fraction of that of their parent bodies. The mass of the Moon is 7.4×10^{22} kg, while that of the Earth is 81 times greater. The mean density of the Moon is 3340 kg/m^3, roughly three-fifths of the average density of the Earth.

There are four main theories concerning the formation of the Moon. The first of these is that the Moon was once part of the Earth but broke away, leaving a hollow which became the Pacific Basin. We now realise that this idea is dynamically untenable. Moreover, the samples of lunar soil brought back to the Earth show that the Moon was most definitely never part of it. Hence, this theory of the formation of the Moon is only of historical interest.

The second idea is that the Moon was formed at roughly the same time as the Earth, about 4600 million years ago, but in another part of the Solar System. The Moon was later captured by the Earth when their orbits crossed. Tidal interaction then caused the Moon to settle in a stable orbit about the Earth. It must be said that there are dynamical difficulties with this theory, though many people have confidence in it.

The third idea is that the Earth and Moon formed together in the same region of space and at the same time. This theory is currently quite popular, though it is not easy to explain the observed compositional differences between the Moon and the Earth. Clearly, we need to develop a better understanding of the processes that must operate within a cloud of primordial matter as the condensation of protoplanets takes place.

The fourth idea, also currently popular, is that at some time in the early history of the Solar System the Earth collided with another large body, perhaps a protoplanet or a large asteroid. Much of the debris thrown up then gathered into an orbit around the Earth and eventually coalesced to form the Moon. Perhaps the Moon was mostly formed from the material that made up the impacting body and at least this might explain the chemical differences between the Moon and the Earth of today.

6.2 The Orbit and Phases of the Moon

The Moon orbits the Earth at a mean distance of 385 000 km, with a sidereal period of 27.3 days. It is commonly known that the Moon exhibits *phases*, where the amount of the Moon we see illuminated by the Sun varies over a 29.5 day cycle. This cycle is known as the *synodic period* of the Moon. Figure 6.1 (*overleaf*) illustrates how this effect is created. The diagram is approximate in that it does not incorporate the fact that the Earth–Moon system is in orbit about the Sun. This factor produces an appreciable change in the direction of the Sun's light during the course of each sidereal period and is the reason why the sidereal and synodic periods are not equal in length. In effect, the Moon has to go a little further than once round the Earth before reaching the same visible phase as before.

At position 1 on the diagram the Moon is new, and its visible appearance is shown in the corresponding figure in the sequence above. The unlit hemisphere is then turned towards the Earth. At position 2 the Moon is at its crescent phase, and at position 3 the Moon is at first quarter. At position 4 the Moon presents a gibbous phase. At position 5 the Moon has its illuminated hemisphere facing the Earth and then appears full. Over the fortnight that the Moon takes to go from position 1 to position 5 it is said to be *waxing*. During the next fortnight the sequence continues, with the Moon shrinking or *waning* in phase. Finally, 29.5 days after new Moon, the Moon is once again new.

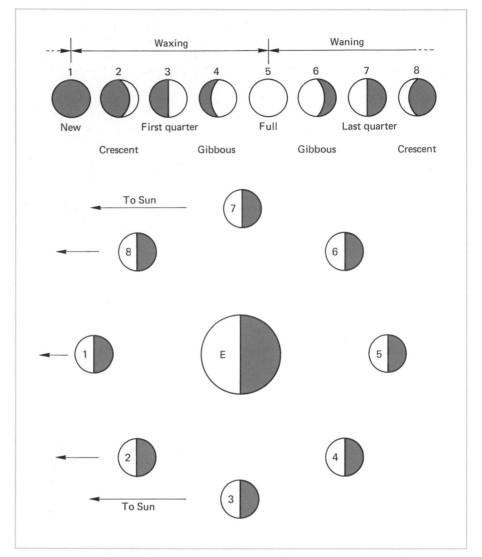

Figure 6.1 The phases of the Moon. The lower part of the diagram schematically represents stages in the Moon's motion about the Earth. The upper part of the diagram shows the corresponding phases as seen from the Earth's surface.

6.3 Lunar Eclipses

The Earth casts a vast, cone-shaped shadow in space. A *lunar eclipse* occurs when the Moon wanders into this shadow. From Fig. 6.1 it might be imagined that a Lunar eclipse occurs every full Moon, but this is not so. The plane of the Moon's orbit about the Earth is inclined at 5° to the plane of the Earth's orbit about the Sun. This means that, for most full Moons, the Moon misses the Earth's shadow-cone and passes above or below it.

If we think in terms of the Moon's apparent path across the celestial sphere, it is a great circle which is inclined at an angle of 5° to the ecliptic. This is illustrated in Fig. 6.2. When the Moon's path crosses the ecliptic, going from north to south, we say that the Moon is at its *descending node*. The other node is the *ascending node*. It should be apparent that an eclipse can occur only when the Moon is at one of its orbital nodes.

Figure 6.3 (*overleaf*) illustrates how a lunar eclipse is formed, though the diagram is drawn very much out of scale for the sake of clarity. First the Moon enters the region of partial shadow, or *penumbra*. If the Moon is not very close to an orbital node, it might well not enter the full shadow at all. Such a situation is known as a *penumbral eclipse*. The amount of dimming of the lunar disk during a penumbral eclipse is almost imperceptible without special measuring equipment.

If the Moon enters the full shadow, or *umbra*, then the direct sunlight will be cut off. A "bite" will slowly appear in the full Moon. If the Moon fully enters the umbra the eclipse is said to be *total*. If only part of the Moon enters the umbra then the eclipse is said to be *partial*. Even at a total lunar eclipse, the Moon does

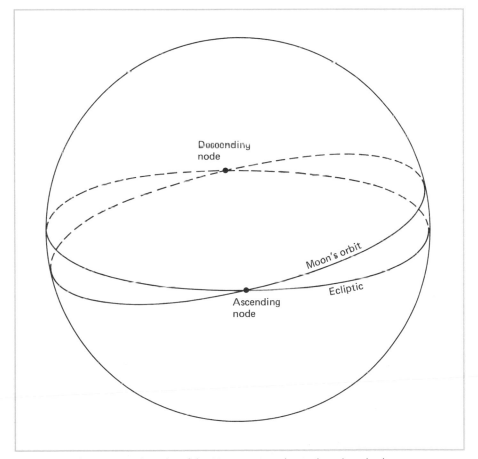

Figure 6.2 The orbit of the Moon projected onto the celestial sphere.

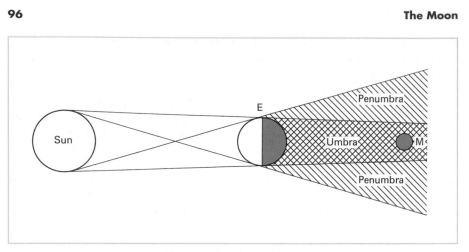

Figure 6.3 A lunar eclipse.

not usually completely vanish from view. Sunlight is refracted and scattered by the Earth's atmosphere, bathing the eclipsed Moon in a reddish glow. Exactly how much dimming there is varies from eclipse to eclipse, depending upon the state of the Earth's atmosphere. During a typical total lunar eclipse the umbra takes about an hour to spread across the face of the Moon. Totality then lasts about an hour, and the umbra takes further hour to leave the Moon. As a very rough average, about two lunar eclipses are visible per year from somewhere on the Earth's surface. A total eclipse of the Moon occurs every few years.

6.4 The Saros

While the Moon moves about the Earth and the Earth–Moon system moves about the Sun, every so often the Earth, Moon and Sun regain the same positions relative to each other. This period, known as the *Saros*, is 6585 days (just over 18 years) long. Thus a lunar eclipse occurring on a particular day will be followed by one 6585 days later. The Saros is less useful for predicting solar eclipses (discussed in Chapter 14), because it is not quite precise enough.

6.5 Libration

Even a casual observer cannot fail to notice that the Moon always keeps the same face presented to the Earth. This is a consequence of the "captured", or *synchronous*, rotation of the Moon. In other words, the Moon takes the same time to turn on its rotation axis as it does to go once round the Earth.

However, the careful observer will notice that the Moon does appear to nod up and down and rock to and fro over a lunar cycle. This effect is termed *libration* and, over a period of time, allows about 59% of the Lunar surface to be mapped. There are three separate causes of libration. *Libration in longitude* is a consequence of Kepler's laws of planetary motion. The rotation rate of the Moon on its axis is constant, but the rate at which it travels around the Earth is not. Consequently, a 7° east–west oscillation is produced over one lunar cycle. Figure 6.4 illustrates libration in longitude.

Libration in latitude is due to a combination of the 5° inclination of the Moon's orbit to the ecliptic and the $1\frac{1}{2}$° tilt of the Moon's rotation axis to the line perpendicular to the plane of its orbit. As the Moon moves around the Earth so it is possible to see, alternately, up to $6\frac{1}{2}$° beyond one pole then the other (see Fig. 6.5, *overleaf*).

Diurnal libration has a minor effect and is caused solely by the rotation of the Earth. This causes the observer's viewpoint to change with respect to the Moon. The effect is illustrated in Fig. 6.6 (*overleaf*). At moonrise the observer can see a little way beyond the eastern limb of the Moon, while at moonset he can see a

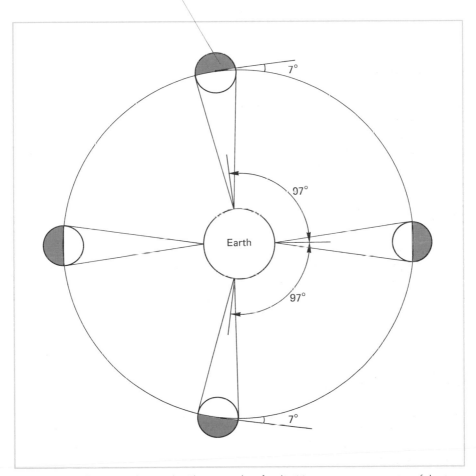

Figure 6.4 Libration in longitude. The time taken for the Moon to go one-quarter of the way around its elliptical orbit in not quite the same as the time it takes to make a one-quarter turn on its spin axis. From the Earth this out-of-step motion causes the Moon to appear to swivel back and forth in an east–west direction.

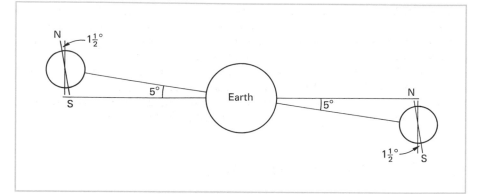

Figure 6.5 Libration in latitude.

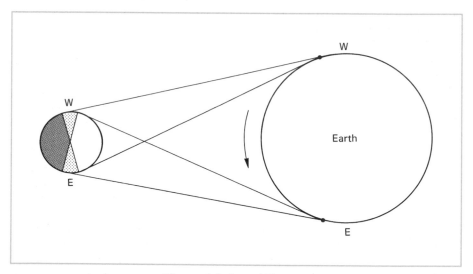

Figure 6.6 Diurnal libration.

little way beyond the western limb. The way in which these librations combine is complicated, and the fact that both the Earth's and the Moon's orbit precess (the positions of the nodes move with time) gives rise to the fact that librations differ with each *lunation*. (A lunation is a complete cycle of lunar phases.)

6.6 The Measurement of the Moon's Distance

By knowing the radius of the Earth and measuring the diurnal libration it is possible to find the distance to the Moon. However, the measurement of diurnal libration is difficult to make with any precision. A better method is to find the mass of the Earth (by measuring the acceleration due to gravity, the radius of the Earth, and the value of G) and then infer the Moon's distance from its orbital period. Even this method is not capable of the highest accuracy, and it has been superseded by direct measurement. Laser pulses are fired at the Moon (on the surface of which special reflectors have been set up by the Apollo astronauts) and the time for the reflection to be received is measured. The distance of the Moon can now be established to an accuracy of few centimetres, which is quite remarkable for a body that is more than a third of a million kilometres away.

6.7 The Selenographers

Even a casual glance will show that the Moon is not a blank, silvery disk. Apart from the phases, patchy markings may be seen. These markings give rise to the supposed outline of a human face, the so-called "Man in the Moon", which may be seen around the time of full Moon.

Galileo, in 1609, used his newly made telescope to observe the Moon and he drew rough sketches of its surface. The large dark patches we now call *maria* (the Latin name for seas), though there are no oceans on the Moon.

Galileo also observed mountains on the Moon and large circular formations that seemed to crowd much of its surface (these circular basins we call *craters*). A modest pair of binoculars will show the Moon as well as Galileo saw it, even with the best of his small refractors.

The fact that the maria (mare is the singular of maria) are darker in hue than the rest of the surface was taken to indicate a difference of chemical composition. We now know this to be correct. In pre-space-age times the dark plains were termed *lunabase*, whilst the light-coloured and rocky highlands were termed *lunarite*.

The system of nomenclature in use today is based on that of Giovanni Riccioli, who published a map of the Moon in 1651. This map was based on observations by his pupil, Francesco Grimaldi. The mountain ranges are named after those on the Earth, such as the Alps and the Apennines. The craters are given the names of famous scientists. The lunar "seas" are given more fanciful names, such as Mare Tranquillitatis (Sea of Tranquillity) and Sinus Iridum (Bay of Rainbows). The nomenclature has been extended to features on the side of the Moon that we cannot observe from the Earth. An outline map of the Earth-facing hemisphere is given in Fig. 6.7 (*overleaf*). It shows features which can be seen by using binoculars or the smallest telescope. Figure 6.8 (*overleaf*) shows a selection of large areas of the Moon's surface as they might be seen through a very small telescope (say a 5 cm refractor, with a magnification of around ×100). Some more detailed views are presented later in this chapter.

When seeing the full Moon high in the winter sky and illuminating the earthly landscape, it is difficult to imagine that it is made up of relatively dark rock. In fact, the average *albedo* of the lunar surface is 0.07. This means that the Moon reflects away only 7% of the sunlight falling upon it.

Largely through amateur efforts, selenography (Moon mapping) advanced considerably after the time of Riccioli, the names of Tobias Mayer, Johann

Figure 6.7 Outline map of the Moon.

Schröter, as well as Wilhelm Beer and Johann Mädler being especially notable. The first photograph of the Moon was made in 1840 by J.W. Draper and photographic atlases soon followed. However, no photograph can equal the detail seen at the eyepiece of a telescope (atmospheric turbulence blurs the image while the exposure is being made), and maps were made that were based upon photographs but with the finest detail added by direct visual observation.

6.8 Lunar Probes and Man on the Moon

The first unmanned lunar probes were launched in 1959 by the Russians and they obtained pictures of the far side, a region never before seen by man. One of the Russian probes, *Lunik 2*, was deliberately crashed onto the Moon. The probe

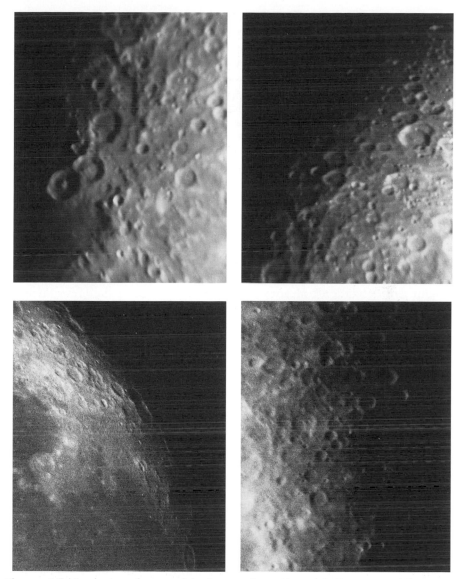

Figure 6.8 A selection of areas of the Lunar surface. *Upper left*: The region of the craters Cyrillus, Catherina and Theophilus. These are the major craters shown and are bordered on their right by the Altai Scarp (Rupes Altai). *Upper right*: Part of the south-eastern quadrant of the Moon. *Lower left*: The south-western border of the Oceanus Procellarum. *Lower right*: A region just south-east of the centre of the Moon's disk. Apart from the upper two, all the photographs were taken on different dates and so show different lighting conditions. Photographs taken by the author, using his $18\frac{1}{4}$ inch reflecting telescope. South is uppermost in these telescopically normal views.

Ranger 7 was the first American vehicle to "land" on the Moon, when it was (quite deliberately) crashed onto the lunar surface in the region of the Mare Nubium on 28 July 1964. A long series of photographs were taken up to a fraction of a second before the probe bit the lunar dust. Many other American and

Russian craft visited the Moon in the years that followed. Lunar mapping culminated with the *Orbiter* series of probes between 1966 and 1968.

While the unmanned probes were being dispatched, preparations were being made to send men to visit our neighbouring world. Major Yuri Gagarin was the first man in space, sent aloft in the Russian *Vostok* capsule, on 12 April 1961. To detail the subsequent development of space flight would require a book on its own, but suffice it to say that the Americans eventually overtook the Russians in what at the time was nicknamed "the space race".

The Christmas of 1968 was marked by the first manned flight to leave the Earth's orbit. The American astronauts Borman, Lovell and Anders, launched in the *Apollo 8* capsule atop a massive, multi-stage Saturn 5 rocket, completed an orbit of the Moon. All who lived through those times will recall the heady excitement which culminated in the *Apollo 11* Moon landing of 20 July 1969. Neil Armstrong became the first man to set foot upon the Moon with the words "That's one small step for man … one giant leap for mankind." Those words have become recorded in history and will be remembered for all time. Edwin "Buzz" Aldrin was the next man out of the spidery looking lunar excursion module (LEM) and onto the lunar surface. He described the scenery around him as a "magnificent desolation". Meanwhile, the third member of the crew, Michael Collins, stayed in lunar orbit in the command module.

Five more Moon landings followed, the last being *Apollo 17* in 1972. There would have been one other, but an explosion aboard the *Apollo 13* command module caused the proposed Moon landing to be aborted and nearly caused the deaths of the three crew members. Originally, the *Apollo* series was to go beyond number 17, but a combination of financial cutbacks and public indifference caused the programme to come to a premature end.

For more than two decades thereafter the Moon returned to being a lonely object for astronomers to peer at through their telescopes but then came *Clementine*. This probe was injected into lunar polar orbit in February 1994 and it was equipped to carry out a complete survey of the Moon in eleven different wavebands spanning ultraviolet through to infrared. Being in a polar orbit, it built up a complete set of overlapping sweeps above the lunar surface as the Moon "turned under it", so to speak. Some of the images obtained were of particularly high resolution (the finest details shown being of the order of 100 metres). There was also a laser-ranging system (called LIDAR) which built up a laser-echo map of even finer resolution.

The data obtained are still being analysed, but it is fair to say that this probe will result in a considerable increase in our knowledge of the Moon's physical, topological and chemical structure. From that scientists might get more clues about the internal make-up of our satellite and its past history.

6.9 The Moon's Appearance

The Moon is the easiest of the celestial bodies to observe. It graces our evening skies for roughly two consecutive weeks every month and it is very close to us on the astronomical scale. A small telescope will reveal a mass of detail and the sight of the lunar surface in a 6 inch (15.2 cm) reflector is impressive indeed.

Using such a telescope with a magnification of, say, ×150, the appearance of the surface tends to remind one of plaster of Paris. The overall colour is a light grey, with the large plains of the maria a darker grey and appearing to be tinted with faint blues and greens (this is an illusion; the real colours of the lunar surface are mostly various shades of brown, but the human eye tends to adapt its colour registration so that the average colour of the Moon comes out as white). Around the time of full Moon the lunar surface is dazzlingly bright and seems to be a mass of streaks, spots and blotches. The sunlight is then striking the lunar surface from virtually the same direction in which we are observing it and so we see no relief caused by the casting of shadows. Around the time of the first quarter and last quarter Moon the effect is not so confusing. Shadows make the lunar features stand out, especially along the *terminator* (the boundary between the lit and unlit hemispheres of the Moon).

Apart from the maria, the features that dominate the lunar scene are the craters. When seen near the terminator the craters are shadow-filled and give the impression of being very deep. In fact, they tend to be rather shallow when compared with their diameters. Craters come in all sizes, from about 300 km in diameter to just a few metres and even smaller. Craters saturate the rough, highland, areas (see Figs 6.9, *below*, and 6.10, *overleaf*). There is, however, a noticeable paucity of large craters on the maria. Our 6 inch reflector is capable of showing craters that are larger than about 2 km in diameter, but it shows many areas of

Figure 6.9 The southern highlands of the Moon imaged by the author, using his 18¼ inch reflecting telescope at the date and time shown. South is uppermost. The crater a little lower left of the centre is called Regiomontanus. Notice the mountain with the summit craterpit within Regiomontanus.

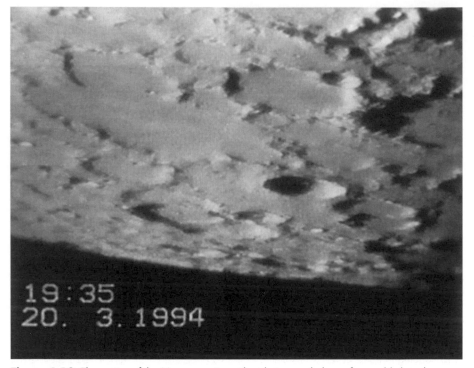

Figure 6.10 The region of the Moon near its north pole is revealed very favourably here because of the unusual libration at the date and time this was taken. Other details as for Figure 6.9.

the mare to be smooth and apparently craterless. The photographs relayed from close-range orbiting probes showed that these areas are in fact saturated with immense numbers of very small craters.

Often the floors of larger craters are strewn with smaller craters and, indeed, there are many examples of craters that break into others. In most cases the smaller crater breaks into the larger. Examples include Clavius, Gassendi, Posidonius, Hevelius and Cavalerius. The number of large craters on the Moon is much less than the number of small craters. Apart from their sizes, craters differ in other respects. Some craters, such as Copernicus, have highly terraced walls. Others have a relatively smooth wall and floor, such as the small craterlets in Clavius. Some craters, including Copernicus and Tycho, have large mountains centrally placed within them. Some others have their floors covered in mare material. Examples of this type are Archimedes (Fig. 6.11) and Plato (Fig. 6.12, *overleaf*). Some craters have had their walls broken down and are nearly submersed in mare material.

For years a controversy raged as to whether the lunar craters were of igneous (volcanic) or impact (meteoritic) origin. Undoubtedly both sorts of craters exist on the Moon, though it is now widely accepted that meteorite impacts have caused most of them.

Some craters have bright interiors and are also the centres of systems of radial streaks of bright material, termed *rays*. The rays are thought to be formed when

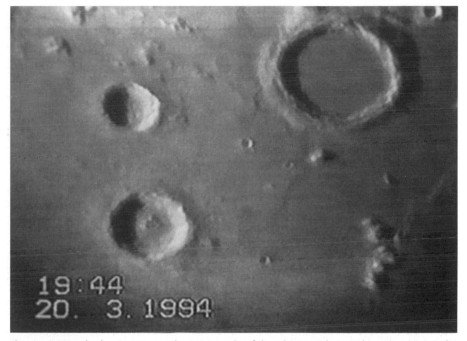

Figure 6.11 The large crater to the upper right of this photograph is Archimedes. Notice the lava flooding that has covered its floor. Of the two large craters to the left, the upper one is named Autolycus and the lower one Aristillus. These are situated on the Moon's Mare Imbrium. Notice the relative paucity of craters on the Mare. Details as for Fig. 6.9.

sub-surface material is splashed across the surface by whatever event forms the crater. The bright material is taken to indicate the crater's relative youthfulness, since it is supposed that the action of the Sun (solar wind and radiation – discussed in Chapter 14) darkens the bright sub-surface material once it has been excavated onto the surface of the Moon. The most notable example of a crater with a bright interior and rays is Tycho, in the southern highlands. This crater is easily visible in binoculars as a bright point and, around the time of full Moon, the rays seem to extend more than half way round the globe. Crater rays are only seen well under a high Sun.

As well as craters, systems of crack-like formations, or *rilles*, are also to be seen on the Moon's surface. Some of the coarser ones, such as the system associated with the craters Triesnecker and Hyginus, are readily visible through a 6 inch telescope. One of the most prominent rilles on the Moon is that in the crater Petavius, running from its central peak to its outer wall. When Petavius is near the terminator the rille is partially shadow-filled and is then easily visible in a 3 inch telescope. There are many examples of much finer rilles all over the surface of the Moon. It is speculated that they were formed from collapsed lava tubes. Hadley Rille was visited by the *Apollo 15* crew, when they landed in the region of the Apennine Mountains (properly called the Montes Apenninus).

Unlike the maria, the lunar highlands are rough and hummocky on a large scale. Mountains and large craters abound. However, mountain chains are

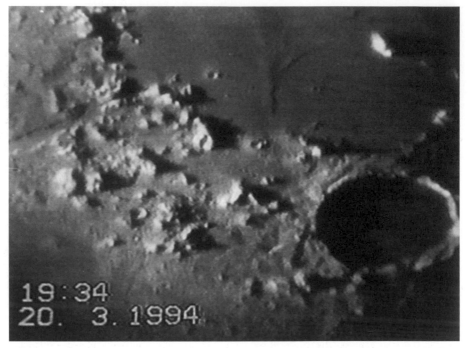

Figure 6.12 The northern extreme of the mountain range named the Montes Alpes which form the north-eastern border of the Mare Imbrium, terminating in the crater Plato (*lower right in this photograph*). Notice the bright peak of the Mons Pico to the upper right, lying on the southern (upper) rim of the faint outline of a "ghost crater" (one buried by lava). Details as for Figure 6.9.

usually to be found bordering the maria but isolated peaks do exist (such as Mons Pico, well shown in Fig. 6.12). Relatively low "swellings" or "blisters" are also seen on the lunar surface. It is thought that these *domes* are produced by ancient volcanic upswellings and some are seen to be topped by small craterlets.

Numerous faults due to horizontal and vertical slippages are seen on the lunar surface. In fact, the Moon is criss-crossed by a pattern of faults and ridges known as the "grid system", consisting of two families of faults running in well-defined directions, roughly perpendicular to one another.

6.10 Surface Conditions

Since the Moon does not appear to travel across the sky at the sidereal rate, it often passes in front of, or *occults*, stars. When it does so, the star does not twinkle and slowly fade from sight but rather suddenly snaps out of sight. In pre-space-age times this was taken to indicate that the Moon has little or no atmosphere. Space missions have confirmed this. Experiments performed by the *Apollo* astronauts have revealed traces of hydrogen, helium, neon and argon, but with incredibly low densities. These gases are thought to be transient and to result from captured solar wind.

Temperatures on the lunar surface vary greatly, because of the absence of a blanketing atmosphere. The equatorial "noon" temperature reaches 100 °C, whilst at night falling to –150 °C! Bolometric observations from the Earth during Lunar eclipses have shown that the surface temperature falls very rapidly once the sunlight has been removed. This indicates that the surface material has a very low coefficient of thermal conductivity – a fact confirmed by the samples brought back from the *Apollo* missions.

The thermal stresses caused by such rapid temperature changes might be expected to erode rocks and crumble the topsoil. Again, this is what was actually found on the lunar surface by the men who went there.

Micrometeorites continually rain down on the lunar surface. Their impacts, together with the thermal effects already mentioned, have tended to crumble and mix the lunar topsoil with slightly deeper layers over the aeons – a process picturesquely known as *gardening*.

6.11 The Structure and Evolution of the Moon

The Russian probes of 1959 were the first to send back photographs of the Moon's reverse side. These photographs were very crude and fuzzy but immediately revealed an important difference in the two hemispheres. Unlike the Earth-facing hemisphere, there is almost a complete absence of maria on the reverse side of the Moon. Instead we find mostly the same sort of rough and crater saturated material as in the highland areas on the near side.

The seismic detectors left on the Moon gave information for years after the last of the *Apollo* missions. From the study of minor tremors in the Moon, *moonquakes*, a picture of the interior structure of the globe has been built up.

The crust extends to a depth of about 60 km and can be divided into three distinct layers. The upper layer is made up of fragmented bedrock. Below this, to a depth of about 20 km, is the lower regolith. This region is composed mainly of basaltic rocks. The lowest crustal layer is mainly composed of rock types known as anorthositic gabbro.

Below the crust is the mantle, consisting of dense rocks rich in the minerals pyroxene and olivine. The mantle becomes progressively less rigid with increasing depth. The presence of a core has not definitely been established from the seismic data, but if one exists then it must be smaller than five or six hundred kilometres in diameter.

The value given for the thickness of the crust is only an average. It is much thinner on the Earth-facing hemisphere, being little more than 20 km in places, whilst increasing to over 100 km on the reverse side. This is thought to explain the difference in the amount of maria-covered surface between the two hemispheres. To understand this let us take a brief look at the evolution of the Moon after its initial formation.

When the Moon was still a molten mass, about 4600 million years ago, differentiation caused the heavier elements to sink whilst the lightest floated to the surface. These lightest materials formed the lunar highlands of today.

During its early life the Moon was bombarded with debris left over from the formation of the Solar System, in common with the rest of the planets. During this period massive lumps of debris smashed into the Moon, creating great basins. As the debris was gradually swept up by the Moon and planets, so the ferocity of the impacts dwindled. By about 3850 million years ago the process was virtually complete, leaving a heavily scarred surface on the Moon. Extensive fracturing of the lunar crust also occurred and low viscosity lava flooded out in successive stages to fill the large impact basins and so form the maria. We see the evidence for these successive lava flows in the form of *wrinkle ridges*, forming the boundaries between the various flows.

Where the crust was thin magma could escape to the surface. Where the crust was thicker it could not. Hence the difference between the two hemispheres. Moreover, the Moon is not a perfectly round globe but bulges a little in the direction of the Earth. The Earth's gravity has worked on this bulge, causing the Moon to attain a *captured*, or *synchronous*, orbit (i.e. the rotation period and orbital period are the same).

Other irregularities occur in the figure of the Moon. Deformations in the paths of orbital spacecraft indicate local increases in density below the lunar surface. These *mascons* are mainly situated below the maria.

As the interior of the Moon cooled, so the flooding activity dwindled and eventually stopped about 3200 million years ago. Small-scale cratering by impact and and volcanic processes continued but by about 1000 million years ago all the major activity had ceased and the Moon was then virtually as we see it today.

6.12 Transient Lunar Phenomena

Perhaps the Moon is not completely inactive even today. Local obscurations and glows, sometimes coloured, are occasionaly seen by regular observers of the lunar surface. These were dismissed as tricks of the eye by most astronomers until 7 November 1958 when Nikolai Kozyrev, using the 50 inch (1.27 m) reflector of the Crimean Astrophysical Observatory, observed a reddish glow and obscuration over the central peak of the crater Alphonsus. He took spectra of the area and these showed that a gaseous emission had taken place.

Many amateur astronomers took up regular monitoring of the lunar surface, and Kozyrev's results also excited some professional interest. Much data has emerged over the years. It seems that certain areas of the Moon are prone to these events, known as *transient lunar phenomena* (TLP), particularly the borders of the maria and certain craters, such as Aristarchus and Plato. Many consider TLPs to be completely illusory, but much evidence has emerged to suggest that they are real. Certainly, though, they are extremely feeble phenomena and probably are caused by no more than gentle gas releases from below the lunar surface (radon gas has certainly been detected during the Apollo missions and the indications are that the sites of maximum emission coincides with the TLP-prone areas). We have learnt much about our neighbouring world, but clearly there are mysteries yet to be solved!

Questions

1 Describe fully the appearance of the Moon in a small telescope, naming the various types of visible features.

2 Describe, using a diagram, how the phases of the Moon arise. Describe, using a diagram, how a lunar eclipse occurs.

3 Explain the term *libration* in connection with the Moon, and show how the various types of libration are produced. What is the *Saros*?

4 Write an essay on the formation of the Moon and its structure, particularly outlining the formation of the maria, craters and the lunar highlands.

5 Briefly outline the surface conditions on the Moon.

6 Find out about, and write an essay on, the main steps that led to man's exploration of the surface of the Moon.

7 The lunar crater Copernicus lies fairly near the centre of the Moon's visible disc as we see it from Earth. It has a mountain mass at its centre. (a) Make a simple sketch of the crater as you think it would appear through a telescope if the Moon's terminator were just a little to the right of the crater. Mark an arrow close to your drawing to indicate the direction of sunlight (the telescope shows south at the top, as is normal for an astronomical telescope set up in the Earth's northern hemisphere). (b) State which of the following is very approximately the actual lunar phase represented by your sketch: New Moon; first quarter Moon; full Moon; last quarter Moon. (c) Describe (or sketch, where appropriate) the appearance of the crater Copernicus through the telescope at each of the other lunar phases given in the list.

Chapter 7

The Inferior Planets

BEFORE the invention of the telescope the only planets known were Mercury, Venus, Mars, Jupiter and Saturn. Their true natures remained unknown, because they appeared like stars to the unaided eye. These planets could only be detected by their brightnesses and their movements relative to the stars. In fact the word planet means "wanderer".

When telescopes were brought into play these wandering stars were revealed in their true guise – whole worlds, in the same way that the Earth is a world.

7.1 Stellar and Planetary Brightnesses

Go out on any clear night and you will see stars of varying brightness. Some are bright while others are barely perceptible. Over the centuries a scale has been evolved to describe the brightness of any celestial body as viewed from the Earth. The scale was at first empirical but has now been given a mathematical grounding.

The number corresponding to a star's brightness on this scale is termed its *apparent magnitude*. The magnitude scale can cause confusion to the uninitiated because the larger positive magnitude corresponds to the fainter star. The faintest stars that can be seen with the naked eye on a clear night of good transparency are of magnitude 6. These stars are then said to be of "the sixth magnitude". The bright star Vega, in the constellation of Lyra, has a magnitude of 0.0, whilst a few of the very brightest stars are brighter still and have negative magnitudes. The brightest star in the sky (apart from the Sun) is Sirius, in the constellation of Canis Major. It has a magnitude of –1.5.

The magnitude scale is logarithmic, with each magnitude corresponding to a brightness difference of 2.512 times. Thus a difference of 5 magnitudes corresponds to a brightness difference of 100 times (2.512 × 2.512 × 2.512 × 2.512 × 2.512 = 100). On this scale, the Sun has an apparent magnitude of –27!

7.2 The Orbit and Phases of Mercury

Two of the Solar System's planets, Mercury and Venus, orbit closer to the Sun than the Earth. They are termed *inferior planets* because of their smaller orbits.

In many ways the innermost planet, Mercury, is a planet which has been loath to give up its secrets. This is mainly due to its close proximity to the Sun. It orbits at a distance of 58 million km, with a sidereal period of 88 days. However, its orbit is highly eccentric and while the perihelic distance is 47 million km, the aphelic distance is 69 million km.

This means that not all apparitions of Mercury are equally favourable. The maximum angular distance between Mercury and the Sun varies from 18° to 27° at *elongations*. Mercury never appears far from the Sun but, as the months go past, appears first to the west of the Sun and then to its east. When Mercury is at its greatest distance to the west of the Sun it is said to be at *western elongation*, and the planet can then be viewed in a twilight sky in the morning where it rises a little before the Sun. At *eastern elongation* Mercury is at its furthest distance to the east of the Sun. It then follows the Sun down in a twilight sky after sunset.

Since Mercury orbits closer to the Sun than the Earth, we see phases, similar to those of the Moon. However, Mercury is unlike the Moon in that it changes its apparent size along with its phases. Venus shows the same effect and the principle is explained in Figure 7.1 (*overleaf*).

At positions 1 and 3 Mercury is at its greatest western and eastern elongations, respectively, and then shows a half- phase, or *dichotomy*. At this point the planet subtends an apparent diameter of about 9 arc seconds. At position 4 Mercury is, to all intents and purposes, between the Earth and the Sun. It then presents its unilluminated hemisphere to the Earth. It is also at its closest to us and so appears at its largest, subtending an apparent diameter of 13 arcseconds, on average. At position 2 Mercury is virtually on the opposite side of the Sun, as viewed from the Earth, and then appears fully illuminated but at its smallest apparent diameter, a mere $4\frac{1}{2}$ arc seconds. Positions 4 and 2 are those of *inferior conjunction* and *superior conjunction*, respectively.

The synodic period of Mercury, that is the period between successive conjunctions, is 116 Earth-days. Figure 7.1 might give the impression that Mercury should be visible as a black spot against the Sun at every inferior conjunction of the planet. In fact, owing to the inclinations of the orbits of Mercury and the Earth, such *transits* are rare.

While the phase of Mercury increases, so its apparent diameter shrinks. This causes the apparent brightness to vary in a complicated way. At its brightest the planet reaches an apparent magnitude of –1.9, a little less than half of the illuminated hemisphere then being presented to us.

7.3 Mercury's Telescopic Appearance

Mercury is a troublesome planet to observe. It is never far from the Sun and when visible in a moderately dark sky it has a low altitude and so is then

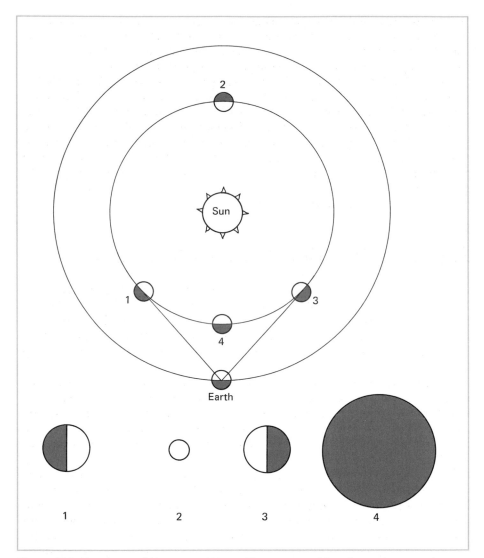

Figure 7.1 The phases of Mercury and Venus. The positions of the inferior planet are shown in the upper part of the diagram, with the corresponding apparent sizes and phases as seen from the Earth shown in the lower part.

seen through a great thickness of the Earth's unsteady atmosphere. Moreover, the planet is very small, having a diameter of 4900 km. A magnifying power of about × 200 is needed at the time of maximum elongations to make its image appear the same size as the Moon seen with the naked eye. Around the times of conjunction the planet is immersed in the glare of the Sun and so is then unobservable.

Most serious observers of Mercury observe it in broad daylight, with the planet high in the sky. Obviously the telescope needs to be fitted with good setting circles in order to find the planet (using a telescope to sweep the sky in the proximity of the Sun is very risky, because the solar disk might accidently enter the field of view with disastrous consequences to the eyesight of the observer) but when the sky is clear and deep blue the planet can be studied.

A 6 inch (15.2 cm) reflector will show dark patches on Mercury's pinkish disk when the conditions are favourable. Probably the best map of the surface of the planet before the space age was that produced by E.M. Antoniadi in 1933. He used the 33 inch (0.83 m) refractor of the Meudon Observatory in France. He had also used this telescope to produce an excellent map of Mars and he was a very experienced observer. However, we now know that Antoniadi's map was very far from accurate. This was not his fault. Mercury is just too difficult to observe from the Earth with any degree of reliability. In fact, the wrong rotation rate was very confidently ascribed to Mercury until as recently as 1965.

7.4 The Rotation Rate of Mercury

From observations of the dark markings, astronomers had long concluded that the rotation period of the planet was the same as its sidereal period, namely 88 days. Thus Mercury was considered to have a captured rotation, rather like the Moon, and would keep one face of the planet constantly facing the Sun. The opposite face would then be permanently exposed to the intense cold of night.

Astronomers measured the radio flux from Mercury in 1962 and were surprised to find that the average flux from the unlit hemisphere was much larger than had been predicted, indicating that it was far warmer than expected. Scientists calculated that the Sunward hemisphere should reach a temperature of 400 °C, while the unlit hemisphere should be at a temperature of no higher than −250 °C.

The temperature of the sunlit hemisphere was confirmed, but the hemisphere experiencing night seemed to be not much colder than 0 °C! Actually, the temperatures measured by the radio techniques were of sub-surface soil, and we now know that the maximum daytime surface temperature reaches 430 °C, whilst the minimum night temperature is −170 °C. The puzzle was solved in 1965 when the 305 m radio telescope at Arecibo was used to measure the rotation period of Mercury by a radar technique.

Basically, a radar pulse was fired a Mercury. Since the planet is rotating, one side of the planet approaches us, while the other recedes. The radio waves arriving on the approaching side of the planet will be compressed as they are reflected back to the Earth. On the other side the waves are stretched (see Fig. 7.2, *overleaf*). Thus the waves arriving back at the Earth will consist of a spread of wavelengths – the amount of spreading depending on the rotation rate. In this way, Mercury was found to turn on its axis at a rate of once every 58.65 days, exactly two-thirds of the planet's sidereal period. This means that the year on Mercury is only $1\frac{1}{2}$ Mercurian days long.

Rather than Mercury turning on its axis once in the same time that it takes to go once round the Sun, three rotations of the planet occur while the planet goes twice round the Sun. This *spin–orbit coupling* is thought not to be coincidental.

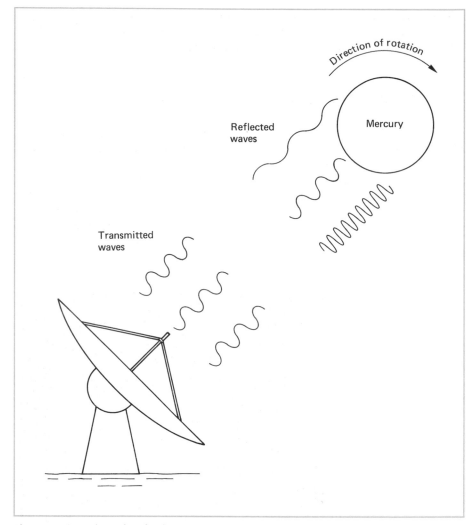

Figure 7.2 Radar pulses fired at Mercury. The spread of wavelengths in the reflected signal gives a measure of the rotation rate of the planet.

Mercury is not perfectly round but is slightly egg-shaped. In other words, Mercury is slightly larger along one diameter than another. Further, at perihelion, when the planet is closest to the Sun, the longest axis of Mercury is aligned to the Sun. At this time the gravitational influence of the Sun on Mercury is at its maximum.

In this way the Sun is responsible for keeping the longest axis of Mercury pointing towards it when the planet is at perihelion. Only certain values of rotation rate are possible if this condition is to be met, and the 3:2 ratio is one of them. The same sort of mechanism operates between the Moon and the Earth, keeping the Moon in its synchronous orbit.

The radar results also showed that Mercury's spin axis is virtually perpendicular to its orbital plane and so the northern and southern hemispheres of the planet do not experience seasons, unlike those of the other planets.

7.5 *Mariner 10*

In April 1974 *Mariner 10* bypassed Mercury, ending centuries of ignorance about the planet. Many instruments were carried by the probe, including a camera that could take and transmit high-quality photographs. The photographs revealed a surface that is strikingly similar to the Moon (see Fig. 7.3). Mercury is heavily cratered and has maria, though these are not as darkly contrasting as those of the Moon. The lack of surface corrosion indicates that Mercury can never have had an extensive atmosphere. However, there are various differences, notably in the interior construction of the two worlds, and so it is as well to compare them in detail.

Figure 7.3 *Mariner 10* photograph of Mercury. JPL photograph.

7.6 Comparison Between Mercury and the Moon

(a) Similarities

Both worlds are heavily cratered, with many bright-rayed craters being visible. Extensive lava flooding also appears to have occurred on Mercury, as the planet shows several large plains relatively devoid of large craters. Both worlds are notable for the absence of an eroding atmosphere. Again like the Moon, Mercury is asymmetrically developed, one side of the planet bulging outwards more than the rest and the dark plains are concentrated on this side.

(b) Differences

Most significant was the discovery of a magnetic field surrounding Mercury. It has a strength about 1% that of the Earth. Together with gravimetric measurements, it indicates that Mercury has a metal core, probably composed of iron, which is a full four-fifths of the diameter of the planet! By contrast, the Moon has no magnetic field and no definitely observed metallic core. The average density of Mercury is 5420 kg/m^3, similar to the density of the Earth and rather larger than that of the Moon.

The craters on Mercury tend to be smaller than those on the Moon. Also, the crater ejecta and secondary-impact craters are much more closely clustered around the primary craters than their lunar counterparts. This effect is due to the Mercurian gravitational field strength at the surface of the planet being about twice as strong as the Moon's field at its surface.

The Mercurian craters are structured differently from those on the Moon. Most Mercurian craters larger than about 14 km in diameter possess terraced interior walls, but interior terracing is only common in lunar craters of diameter greater than about 50 km. Several of the large, flat-floored Mercurian craters show complex pattern of ridges on their floors, while the lava-flooded lunar craters have relatively smooth floors.

The surface of Mercury is scarred by huge cliffs, termed *lobate scarps*, that run for hundreds of kilometres. Typically, they reach heights of around 3 km above the surface and they cut across craters and intercrater areas. It has been suggested that they were caused by compressional stresses in the crust as the core cooled after its formation.

Colorimetric and photometric measurements from Mariner 10 indicate that the surface is mainly composed of silicates, with much less iron and titanium than on the lunar surface. These elements are responsible for the relative darkness of the lunar maria, as compared with the highland areas. Their scarcity on Mercury explains the absence of strong contrasts between the Mercurian highlands and the inter-crater plains. Nonetheless, the average albedo of Mercury is very similar to that of the Moon, at 0.06.

7.7 The Caloris Basin

A single feature that dominates the surface of Mercury is the Caloris Basin. This is a huge circular depression, about 1300 km in diameter, bordered by high mountain chains and with large interior concentric rings and a vast array of complicated topography. It is thought that this feature was produced by a massive meteorite impact roughly 3800 million years ago.

There are indications that a wave of global volcanic activity started after the impact. On the opposite side of the globe of the planet there is much evidence of turmoil caused by the shock waves focused from the site of the impact. We can be sure that the event that marked the formation of the Caloris Basin must have been spectacular by any standards!

7.8 The Surface Conditions on Mercury

The surface of Mercury would certainly be a very hostile place for humans. It has a virtually non-existent atmosphere, basically composed of neon, argon, helium and hydrogen, which is thought to be transient and originating from captured solar wind particles. The Sun would be above the horizon for roughly one Earth month, with temperatures hot enough to melt lead. At night an astronaut might be warmed by heat radiating from the ground, but the temperature of the surface would ultimately fall to around –170 °C just before the Sun rose once more to roast the surface in its searing heat. The regions near the poles of the planet experience less extreme temperature variations.

7.9 The Orbit and Phases of Venus

Of all the bodies in the Solar System the planet Venus is probably the most startling to the naked eye. This planet can often be seen shining in the pre-dawn or evening twilight sky with a silvery brilliance that sets it apart from the other planets.

Venus is second in order of distance from the Sun and, like Mercury, it is termed an inferior planet. Venus moves in a nearly perfectly circular orbit of radius 108 million km. It shows phases for the same reason as Mercury. At superior conjunction the distance of Venus from the Earth is at its maximum and the disk of the planet subtends an apparent angular diameter of only $9\frac{1}{2}$ arc seconds. At this time the illuminated hemisphere is turned towards us but the planet is swamped in the glare of the Sun.

As the phase shrinks, so the apparent diameter of the planet increases to reach 65 arc seconds when at inferior conjunction. The planet is then only 40 million km from the Earth, the closest any major planet approaches. At dichotomy, Venus subtends an angular diameter of 25 arc seconds. When at its greatest brilliance, Venus reaches an apparent magnitude of –4.4, much brighter than any object in the sky, apart from the Sun and the Moon. Its angular diameter is then 35 arc seconds.

Venus takes 224.7 Earth-days to go once round the Sun (the sidereal period) and the time that elapses between successive superior or inferior conjunctions (the synodic period) is 584 Earth-days.

7.10 Venus's Telescopic Appearance

Like Mercury, Venus is never very far from the Sun, as seen from Earth. At best it sets six hours after the Sun or rises six hours before it. Thus it is seldom seen high up in a dark sky and serious observers often study the planet in broad daylight.

The impressiveness of the naked-eye view of the planet tends to lead to great expectations of its appearance through a telescope, but the first-time viewer never fails to be disappointed by what is little more than a slightly yellowish-white disk, showing only the phase with any clarity. The reason is that Venus is completely swathed in opaque clouds which block any direct view of the surface of the planet.

The cloud tops of Venus have an albedo of 0.59. That is, they reflect away fully 59% of the sunlight falling upon them. It is because of this that the planet appears so brilliant as seen from Earth. The experienced observer, equipped with a moderate telescope, can sometimes make out a few shady markings and occasional bright spots, but these are all of a very transient nature. However, two anomalous effects are fairly well documented–the so-called "ashen light" and "Schröter's effect".

The ashen light is only rarely seen. When it is, it takes the form of a faint glow illuminating the unlit hemisphere, reportedly often showing a faint brownish or violet tint.

A practical investigation was organised from the Los Alamos Laboratory for the period centred on the June 1988 inferior conjunction of the planet. Observers from around the world were enlisted and the data were compared with the simultaneous *Pioneer Venus Orbiter* spacecraft. The ashen light was seen a number of times simultaneously by a number of widely separated observers. The final report concluded that the effect is real, not illusory, the most definite sighting being my own made from the Royal Greenwich Observatory, and confirmed by another observer on the same site. On the same night – 23 April 1988 – various other observers saw the phenomenon from different locations. I saw the glow as a uniform grey in colour through a 7 inch (178 mm) refractor and 30 inch (760 mm) and 36 inch (915 mm) reflectors and the appearance reminded me of a faint earthshine on a crescent Moon.

The effect has not been explained, though there has been some speculation about a mechanism involving auroral emission. It is true that Venus, being nearer to the Sun than we are, experiences a greater intensity of solar wind, but we also know that Venus has no appreciable magnetic field to concentrate the particles.

Schröter's effect appears as a slight reduction of the phase of Venus. Its effect is to cause the apparent dichotomy to be several days late when the phase is waxing and several days early when the phase is waning. The effect is believed to be caused by the refraction and scattering of Sunlight in Venus's atmosphere, which intensifies the darkening along the terminator. A predictive computer model produces a close match with the observed effect.

7.11 The Size and Nature of Venus

Until recently man was in virtual ignorance of the nature of the surface of Venus. Indeed, until the radar measures of the 1960s the rate of rotation of the planet was not known. It was then found to turn on its axis once every 243 Earth-days, making Venus's day longer than its year! Even more surprising is the fact that the rotation is retrograde. In other words, Venus spins in the opposite direction to the majority of the planets in the Solar System.

As to how this remarkable situation arose, no-one is quite sure. The answer undoubtably lies in the early evolution of the Solar System, about which our ideas are, to say the least, sketchy. Only the planet Uranus, far out in the Solar System, has such an axial inclination that it can be considered to have a retrograde rotation, but Venus's rotation axis is close to being perpendicular to its orbital plane.

That Venus is a rocky world only slightly smaller than the Earth is now well known. Its diameter is 12 100 km, only 700 km smaller than the Earth. Venus has a mass about 81% that of the Earth and a density of 5250 kg/m^3, slightly less than that of the Earth. It is thought to have a core of roughly similar size and composition to that of the Earth. Venus used to be called "the Earth's sister planet" because of the similarity of the physical dimensions of the two worlds, but this term has had little use since we learned more about the surface conditions of the planet.

7.12 Spacecraft to Venus

The years leading up to the first space probe visitation of Venus were rife with speculation about what the probes would find. For instance, ground-based spectroscopic studies had shown that the atmosphere of Venus contained a great deal of the gas carbon dioxide, but water vapour is much more difficult to identify and the evidence as to the amount present was often conflicting. Some people thought that Venus might be largely ocean covered. Some imagined a tropical paradise of Amazon-type swamps and lush vegetation. Some even speculated that atmospheric carbon dioxide would dissolve in the Venusian seas to form oceans of soda-water! Others thought that the surface would be found to be totally dry and desolate, and we know now that this view is the closest to reality.

A giant leap forward was made in 1962 when the American space probe *Mariner 2* flew past the planet at close range. Infrared instruments carried by the spacecraft indicated that the temperature at the surface of the planet is in excess of 300 °C. This result shocked the scientific world and was viewed with scepticism by many. *Mariner 2* also indicated that the planet had no measurable magnetic field.

Mariner 2 was followed in 1965 by *Mariner 5*. This probe confirmed the earlier findings and provided more reliable information. In the same year the Russians managed to soft-land a probe, followed by two more in 1969. Little useful information was obtained from these early Russian probes, as they failed either during descent, or on touchdown. The situation improved in 1970 when the Soviet space probe *Venus 7* parachuted to a soft landing and measured the atmospheric pressure as 91 times that at the surface of the Earth. The temperature proved to be even higher than the *Mariner* probes had indicated – a staggering 450 °C to 500 °C! The probe functioned under the harsh conditions for only 23 minutes.

The next major step in the study of Venus came in 1974 when *Mariner 10* bypassed the planet on its Mercury–Venus mission. This probe took thousands of high-quality photographs of Venus's cloudy mantle in ultraviolet light (see Fig 7.4). These showed the structure and motions of the clouds, and how they swirl down from the planet's poles and toward the equatorial regions. The atmosphere, as a whole, has a rotation period of about four and a half days. *Mariner 10*'s onboard spectroscopic equipment provided information on the composition of the upper atmospheric layers.

These results, together with those of the preceding probes, give us a good idea of the structure and composition of the atmosphere. It is mainly composed of carbon dioxide and extends to a height of about 135 km. The clouds are chiefly composed of droplets of dilute sulphuric acid, stratified into three distinct layers between 47 km and 70 km above the surface of the planet. In addition, layers of sulphuric acid haze extend from below the clouds to a height of 90 km above the surface.

Figure 7.4 *Mariner 10* photograph of Venus, taken in ultraviolet light. JPL photograph.

Of all the spacecraft that visited Venus before the 1980s, only two sent back pictures of the surface – the Russian landers *Venera 9* and *10*, in October 1975. Each took one photograph before succumbing to the conditions. The 1982 probes, *Venera 13* and *14*, also obtained pictures of the surface. Each landing site consisted of a flattish landscape of grey slab-like rocks and sandy aggregates, bathed in the glow from an orange sky.

In 1978 the Americans launched their *Pioneer Venus Mission*. This consisted of an orbiting probe and four landers. The orbiter has allowed the surface of the planet to be radar-mapped to a far finer resolution than is possible from the Earth. Apart from a few elevated features (such as the areas named Ishtar Terra, Maxwell Montes, Aphrodite Montes and Beta Regio), the surface of Venus is remarkably uniform in level. It is thought that no plate tectonics are in operation on Venus and that the crust is very thick.

7.13 Venus – A Volcanic Planet?

Evidence from the *Pioneer Venus Mission* suggested that Venus might support active vulcanism. Since Venus is, effectively, a "one-plate" planet, vulcanism is the only possible mechanism for venting the buildup of heat caused by the radioactive decay of elements in the interior of the planet. The areas suspected of vulcanism certainly look like shield volcanoes and orbital data gave peliminary indications of gravitational anomalies over these areas which were thought to be the result of crustal stressing.

Perhaps most convincing has been the detection of lightning concentrated over two of the suspected areas – Beta Regio and Aphrodite. The lightning is revealed by the production of radio "whistlers", a very distinctive signature. Lightning is known to be associated with terrestrial vulcanism, being caused by friction within the turbulent gases and dust rising from eruptions. However, the situation was far from being settled and to obtain more data another probe was sent.

7.14 *Magellan* – and Yet More Mysteries

Originally a space probe named *VOIR* (Venus Orbiting Imaging Radar) was planned, but heavy financial cutbacks imposed on NASA resulted in a much cheaper probe called *Magellan* to be hoisted aloft by the Space Shuttle *Atlantis* on 4 May 1989. The attached booster rockets then fired to send the probe on a 15 month cruise to Venus. Entering a polar orbit above Venus on 10 August, about a week later it began to fire radar-ranging signals through the clouds and down to the planet's surface to map 25 km wide stripes at very high resolutions (around 120 metres, ten times better than before) and in 3-D. As Venus turned under it, so it gradually built up a map of the whole of the planet's surface.

Many big surprises were beamed back by *Magellan*; among the features of a completely alien landscape, are vast, river-like channels formed not by water but by an obviously very fluid lava (now solidified), huge and steep mountain belts; and steep-sided, dome-like swellings, to mention only a few from a long list.

One of the biggest surprises was that there was no evidence for any significant active volcanism – and no other obvious way for the planet to vent its internal heat. *Magellan* was expected to resolve this issue. Though it failed to provide a definite answer, perhaps a clue does come from one of the other great surprises – the morphology and distribution of the craters.

Studies of the 936 recorded impact craters show that they have a totally random size and distribution. This is a unique situation in the Solar System. Also, they are very obviously not eroded, and so are relatively young – perhaps about 500 million years in age. Putting these facts together leads to the conclusion that the surface of Venus, itself, is only about 500 million years old!

Somehow the planet has completely resurfaced itself about 500 million years ago and yet has not changed very much since that time. Don Turcotte of Cornell University voiced a theory that would explain this and also account for the escape of the heat from deep within the planet. He suggested that the sub-surface heat builds up and gradually melts the mantle of the planet over a period of time. When the melting process reaches it, the surface melts and collapses into the lower magmas, mixing with them. Once the excess heat has escaped the surface cools and solidifies once more and Venus's surface once again becomes quiescent. The process then starts all over again. In other words, the surface of Venus acts rather like a cyclic heat-valve.

This theory met widespread derision, as most planetary scientists prefer gradual to cataclysmic processes. However, they have yet to come up with any other explanation that fits the observed facts.

On 12 October 1989 *Magellan*, its work done, ploughed into Venus's corrosive atmosphere and came to a fiery end. The mission may have been greatly slimmed down from its original concept, but it was still the most remarkable and successful planetary mapping probe to date. The collossal amount of data it sent back will still be being analysed in years to come. *Magellan* has uncovered many more mysteries than it has answered old ones.

7.15 The Earth and Venus – Worlds Apart

Despite the fact that Venus and the Earth are so similar in size and mass, and although they both orbit in the same region of the Solar System, the two planets exhibit remarkably different surface conditions. Venus is not as geologically active as the Earth. Two-thirds of the surface of the Earth is covered in water. Overlaying the land masses and oceans is a cool oxygen–nitrogen atmosphere. Further, the Earth is swarming with a variety of plant and animal life.

By contrast, Venus is a veritable hell! Temperatures on its surface range about 480 °C, with little variation between the night and day sides of the planet. The atmosphere is a soupy, carbon dioxide rich, mixture (roughly 96% carbon dioxide and $3\frac{1}{2}$% nitrogen, with traces of other gases such as hydrogen sulphide). The atmospheric pressure at the surface of Venus is 91 times that at the Earth's surface. Moreover, sulphuric acid contaminants extend from the high haze layers down to the relatively clear (but highly refractive) "air" at ground level. It even rains sulphuric acid on Venus, though the high temperature causes the droplets to be vaporised before they can reach the ground.

Clearly no Earth-like life can have evolved on Venus, and extreme forms of protection would have to be developed before man could ever set foot on its roasting surface.

7.16 The Greenhouse Effect – a Difference in Evolution

The lack of water vapour in the Venusian atmosphere has long been a puzzle. It has always been supposed that both Venus and the Earth formed near their present, close, orbits. Both planets appear to possess similar quantities of carbon and nitrogen. One might suppose that they should have similar quantities of hydrogen and oxygen (the constituents of water) but this is not the case. Free hydrogen and oxygen, as well as water, are extremely scarce on Venus.

Various theories have been proposed to explain the difference but the most likely one stems from the fact that Venus orbits a little closer to the Sun than does the Earth. We think that the atmospheres of both planets initially contained most of the free carbon dioxide possessed by these bodies, also that both planets were largely covered by ocean soon after their formation – certainly, astrophysical calculations indicate that the early Sun was not as luminous as it is today, and so an early Venusian ocean is allowable.

The primitive atmospheres of the two worlds would absorb a portion of the infrared radiation (radiant heat) from the Sun causing their surface temperatures to be somewhat higher than if they had no atmospheres. This is the *greenhouse effect*. As the Sun steadily increased its radiance over the aeons, so a point was reached where the oceans of Venus were partially vaporised. Water vapour strongly absorbs infrared radiation and so the temperature on Venus increased even more. The greenhouse effect became runaway and some people think that the surface of Venus became as hot as 1200 °C, melting the surface of the planet and exposing hot lavas to the atmosphere.

The lavas then reduced the water vapour to a mixture of hydrogen and oxygen, most of the oxygen being chemically combined with the lava and taken down below the surface of the planet with its churning motions. Most of the hydrogen leaked away into space and the planet settled down to its present furnace-like conditions. If this scenario is correct, an enrichment of the hydrogen isotope deuterium should occur in the atmosphere and this has, indeed, been found by the *Pioneer Venus Orbiter*.

While the oceans of Venus were vaporising, water persisted on the more distant Earth long enough for microbial life to develop in the oceans. These microbes began consuming the atmospheric carbon dioxide and releasing oxygen, so preventing the greenhouse effect from becoming runaway as it did on Venus. Having gained a foothold, the presence of life on our world modified the conditions on it, allowing further life to develop. The primordial carbon dioxide became "fixed" in solid carbonates, particularly chalk. Oxygen-breathing animals appeared and evolved and, eventually, man arrived to marvel at the fragile chain of events that led to his existence.

Questions

1 Explain the terms *inferior planet, elongation, superior conjunction,* and *inferior conjunction.* Explain what the *synodic period* is and give its value for Mercury and for Venus.

2 Explain why the planets Mercury and Venus show lunar-like phases and why their apparent sizes vary.

3 Describe the telescopic appearance of the planet Mercury and explain, as fully as you can, why it is a difficult planet to observe.

4 Explain the early confusion over Mercury's rotation rate and how it was eventually resolved.

5 Compare and contrast the planet Mercury with the Earth's Moon.

6 The planet Venus is often known as the "Morning Star" or the "Evening Star". Why do you think this is so?

7 Outline the main points of similarity between the Earth and the planet Venus.

8 Write an essay on the appearance and orbit of the planet Venus. Your answer should contain the varying sizes of the planet and mention of dichotomy and Schröter's effect, etc.

9 Outline the main differences between the Earth and the planet Venus.

10 Describe, in as much detail as you can, the atmosphere of the planet Venus.

11 Briefly outline the major methods of observation of the planet Venus (including space probe) and describe the major facts learned by each of these methods.

12 What is the *greenhouse effect*? How has it caused major differences between the conditions on the Earth and the planet Venus?

13 Find out about, and write an essay on, the *Magellan* space probe mission to Venus and the results and discoveries it has provided – and the mysteries it has uncovered.

Chapter 8

The Planet Mars

BEYOND the Earth's orbit, at a mean distance of 228 million km from the Sun, lies the orbit of the planet Mars. Mars is thus the first of the *superior planets* (planets which orbit the Sun at greater distances from it than the Earth does). Apart from the orbits of Mercury and Pluto, Mars's orbit has the highest eccentricity of all the major planets. The orbital radius of Mars varies from 250 million km at aphelion, decreasing to 208 million km at perihelion. The sidereal period of Mars (the time taken for Mars to go once round the Sun) is 687 Earth-days.

Mars is a small, rocky, world with an equatorial diameter of 6800 km and a mass of only about one-ninth of that of the Earth. Its axial rotation period is 24 hours 37 minutes, very similar to that of the Earth. The inclination of Mars's rotation axis is 24°, again a similar value to that of the Earth ($23\frac{1}{2}$°).

The size and mass of Mars give it a mean density of 3940 kg/m^3, rather less than that of the Earth (5520 kg/m^3). Nevertheless, Mars is a solid, rocky, planet which, like all the planets, emits no light of its own but relies upon the Sun for its illumination.

8.1 Oppositions

Being a superior planet, Mars does not show phases like Venus or Mercury, and usually appears very nearly "full". However, at extremes Mars can display a phase rather like the Moon two or three days before full. Such a phase is known as *gibbous*.

The situation for observing Mars is very much better than for the inferior planets. When Mars is best placed for observation it is closest to the Earth and is on the opposite side of the Earth to the Sun (see Fig. 8.1, *overleaf*). Mars is then seen at its highest in a midnight sky. The planet is then said to be at *opposition*.

Since Mars and the Earth both orbit the Sun in the same directions, the *synodic period* of Mars (the time interval between successive oppositions) is longer than the sidereal period. On average, the synodic period is 780 Earth-days, though subject to large variations. This is due to the non-circularity of the orbits of the Earth and, particularly, Mars causing the orbital speed of the planets to vary as each one goes round the Sun (see Chapter 5).

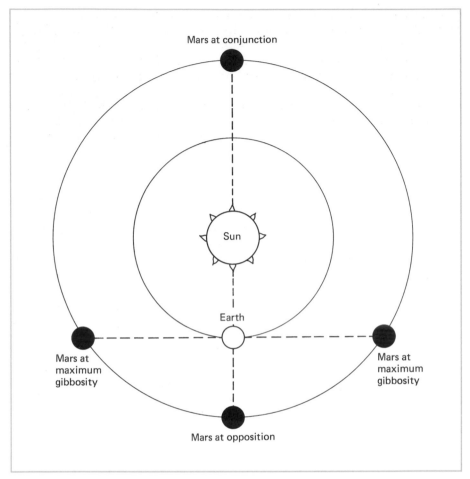

Figure 8.1 Mars at opposition and conjunction.

As Fig. 8.1 shows, about 390 days after opposition the planet will be very close to the Sun, as viewed from the Earth, and it is then said to be in *conjunction* with the Sun. Owing to the ellipticity of the orbits of the Earth and Mars (mainly Mars), not all the oppositions of the planet are equally favourable, the minimum distance of the planet from the Earth varying between 56 million km and 101 million km. At conjunction the Earth–Mars distance increases to around 400 million km.

8.2 The Path of Mars Across the Celestial Sphere

Unlike the planets Mercury and Venus, which appear (from the Earth) to move in and out from the Sun in the manner of cosmic pendulums, the planet Mars is carried round the sky with the constellations (remember it is the motion of the Earth about the Sun which causes the seasonal march of the constellations around the sky).

Also, because of its own motions, the position that Mars takes when it comes to opposition itself moves through the constellations, taking roughly 16 years to go once round the zodiac.

However, the motion of Mars is not always steady and unchanging. Mars moves slowly when near conjunction and moves rapidly, with an obvious nightly shift through the stars, around the time of opposition. Occasionally Mars even stops its "forward", or *direct*, motion through the starry patterns and, for a few weeks, appears to move backwards, or *retrograde*, before once again continuing its direct motion!

This retrograde looping effect is not hard to understand and is simply due to the difference in orbital velocities of the Earth and the planet Mars. Since the Earth moves around the Sun at a mean speed of 30 km/s, whilst the planet Mars moves round at only 24 km/s, the Earth (around the time of Mars's opposition) appears to "catch up" and overtake Mars. Then, for a short period of time while the planets are close, Mars appears to move backwards. Figure 8.2 (*overleaf*) illustrates the principle.

The positions when Mars briefly stops to reverse its direction are known as the *stationary points* in its path across the sky. The other superior planets also show this retrograding effect but to a much lesser extent because of the much greater distances separating these planets from the Earth.

Before we leave the details of the orbit of the planet Mars, it is interesting to note that it was a happy coincidence that Kepler studied Tycho Brahe's observations of Mars, rather than those of another planet, because of its eccentric orbit. If he had chosen, say, Venus (which has a nearly circular orbit) then he may never have discovered the true elliptical figure of the planetary orbits.

8.3 The Telescopic Appearance of Mars

Mars, when near opposition, is a brilliant object in the night sky. Its maximum brightness is magnitude –2.8, so that it can shine more brightly than Jupiter, though usually Jupiter is the brighter of the two. Mars's strong red colour makes it especially distinctive and, indeed, it was its colour that led the planet to be associated with the Roman god of war.

The planet is a difficult object for observation owing to the smallness of its image, even at favourable oppositions. At a favourable opposition (such as that in 1986) Mars can subtend an angular diameter of as much as 25 arc seconds. A magnification of about ×80 will then produce a view of the planet similar in size to the full Moon, when the latter is seen with the unaided eye. At unfavourable oppositions the disk can be less than 14 arc seconds in diameter, and at such times a magnification of about ×140 is needed to enlarge the disk to the same extent as before.

The maximum diameter of Mars at the opposition of 1980 was only 13.8 arc seconds. Moreover, the planet is only at its largest for a few weeks around the

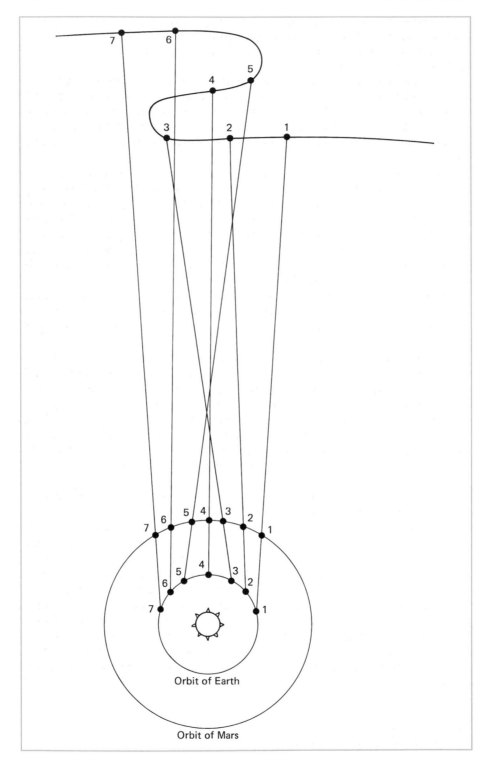

Orbit of Earth

Orbit of Mars

time of opposition (and remember successive oppositions are roughly 2 years and 2 months apart). At the time of conjunction the disk of the planet shrinks to about 3.5 arc seconds, when, in any case, the planet appears too close to the Sun in the sky to be observable.

Remembering that even on fine nights the atmospheric turbulence may not allow detail finer than about 1 arc second to be seen, the problems involved in the Earth-based study of Mars are evident. Scientists had to wait until the space age in order to learn much about the surface of the planet and, indeed, some of the previously most authoritative ideas about the planet proved to be entirely wrong!

The most striking feature of Mars, as seen through any telescope, is its strong reddish-ochre hue. This colour covers most of the visible surface and was, originally, thought to be due to enormous deserts. Telescopes of aperture larger than about 3 inches (76 mm) show darkish markings on the surface of the planet (see Fig. 8.3, *overleaf*). These markings are broadish, resembling islands or continents, and are a different colour from the rest of the surface. Early observers were fooled into thinking that the markings were green in colour (an illusion caused by contrast with the adjacent reddish areas) and attributed them to vegetation.

I find that through a small telescope the markings on Mars tend to appear blue–green in colour, but through larger telescopes the markings appear grey–green or grey (as well as being better seen because of the increased "grey-scale" contrast).

Many early observers also reported seasonal changes in these markings (the Martian seasons being roughly twice as long as those on the Earth). In the hemisphere experiencing Martian winter the dark markings were reported as pale and brownish in tint, but in the Martian summer a "wave of darkening" was observed that swept from the polar to the equatorial regions, when the markings assumed their summer shade of green! This was confidently taken to indicate the seasonal growth and dying away of vegetation, though we now know such observations to be mistaken and no more than wishful thinking on the part of astronomers.

The overall albedo of the surface is 0.15, which means that it is roughly twice as reflective as the surface of the Moon or the planet Mercury.

Features of the planet which do show real variations in accordance with the Martian seasons are the polar caps. Depending upon the tilt of the planet (and hence which pole is presented to us) either one, or both of the polar caps are visible. In a small or moderate telescope they show up as glistening areas of whiteness and they vary their shape and extent as the Martian seasons progress. When the northern hemisphere of the planet is experiencing winter the northern polar cap is at its fullest extent, covering the ground down to a latitude of about 60°, whilst the southern cap is very small and irregular. When the southern hemisphere experiences winter the situation is reversed. We now know that the Martian polar caps are composed of a mixture of water–ice and frozen carbon dioxide.

Figure 8.2 The retrograde motion of Mars.

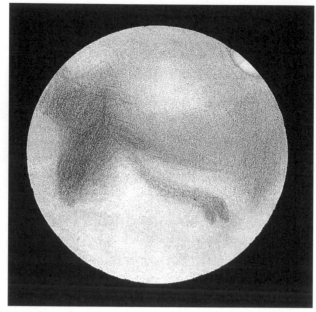

Figure 8.3 Mars as seen through a telescope. Observation by the author on 1988 October 28^d 00^h 20^m UT with the 36-inch (0.9 m) Cassegrain reflector formerly of the Royal Greenwich Observatory at Herstmonceux. Magnification: ×312. Note the bright icecap to the upper right in this drawing (south uppermost) and the large, dark formation of the Syrtis Major to the left.

8.4 The Martian Atmosphere

We have long realised that there is an atmosphere on Mars, but until the space age its composition and extent were not definitely known. Since the acceleration due to gravity (see Chapter 5) on Mars is roughly two-fifths of that of the Earth, it was reasoned that Mars's atmosphere ought to be fairly dense, though the pressure at ground level was thought to be less than that of the Earth's atmosphere at ground level.

Oxygen and nitrogen were prime candidates for the constituents of the Martian atmosphere, with the pressure perhaps high enough to allow areas of liquid water on the surface of the planet. However, the atmosphere was found to be chiefly composed of carbon dioxide and the ground-level pressure turned out to be less than 1% of that at the Earth's surface.

White clouds, composed of ice crystals, are often seen in the Martian atmosphere, and they can take on very large proportions, even covering whole quadrants of the planet. Further, great dust storms are sometimes seen on Mars and, indeed, in a major storm the whole of the surface of the planet can become obscured – Mars then taking on the appearance of a blank, yellowish disk. It appeared like this when it was at its favourable opposition of 1971, the dust storm persisting for several weeks.

8.5 The Canals of Mars

One of the most famous arguments that raged amongst experts of Mars was about the "canals". The controversy originated in 1877 and the last vestiges of it were not finally laid to rest until 1965!

Ever since the invention of the telescope Mars has been a subject for observation. The early drawings of the planet were crude and lacking in detail simply because of the state of the art of telescope manufacture. However, the famous lunar observers Beer and Mädler observed the planet from 1830 to about 1841 and they found that the dark areas were virtually permanent in outline. They even succeeded in producing a map of the planet, using their $3\frac{3}{4}$ inch (94 mm) refractor telescope. Later astronomers, with more powerful telescopes, produced better maps and the study of the Mars continued to advance.

1877 was, perhaps, one of the most memorable years in the history of the study of the Mars. The opposition of the planet, in the September of that year, was a particularly good one. Indeed, it was in this very year that Mars's two satellites were discovered.

Giovanni Schiaparelli used the $8\frac{3}{4}$ inch (221 mm) refractor of the Brera Observatory in Milan to study the planet and he produced a map which was better than any of its predecessors. He replaced the older nomenclature with his own and it is true to say that his nomenclature, in a revised form, survives today.

The remarkable feature of Schiaparelli's drawings is that they show many fine lines across the surface of the planet, joining up the larger dark patches. The lines are either straight or gently curved and they seem to intersect at small, dark patches. In all Schiaparelli observed and mapped about forty of these channels in 1877. Schiaparelli called them "canali" (channels) but, almost inevitably, the name canals caught on.

Many astronomers were sceptical of the existence of these canals but in the following (less favourable) opposition in 1879 many astronomers also saw them. Schiaparelli himself recovered all his old canals and added further ones. He even found that many of his canals were doubled into two running parallel to each other.

It is particularly significant that different astronomers, observing at around the same time, showed completely different arrangements of canals and that none were seen before 1877. It seems likely that few observers would have recorded canals if they had not heard of their "existence" beforehand! This is not to question the honesty of the observers, just the effects of human subjectivity – it is very easy to convince oneself of the apparent reality of such features as faint streaks and spots on an object as difficult to observe as the planet Mars.

The wealthy businessman-turned-astronomer Percival Lowell went so far as to found an observatory in Arizona in order to study Mars and its canals. He equipped the observatory with a 24 inch (610 mm) refracting telescope, which he usually preferred to stop down (using an iris infront of the object glass) to an aperture of 16 inches (407 mm) and set about recording even more artificial looking canals than had Schiaparelli.

Indeed, Lowell even went so far as to postulate that the canals actually were waterways laid by Martians to carry water from the snowy poles of the planet in order to irrigate the desert regions and so produce the vegetation on which their lives depended; needless to say, these ideas were considered controversial in the extreme.

The great observer E.M. Antoniadi studied Mars in the 1920s with the 33 inch (830 mm) refractor of the Meudon Observatory in France and he considered the canals to be illusory. At most he found unconnected spots and small streaks, and he reasoned that the canals were formed by the unconscious tendency of the observer to join them up into straight lines. We now know that Antoniadi was mostly correct, though many of the reported canals do not coincide at all with any albedo features. They were the result of imaginative brains behind eyes that were straining to work beyond their normal limits.

The requiem for the canals finally came in 1965 when the space probe *Mariner 4* bypassed Mars and sent back twenty-one moderately clear pictures of a limited area of the planet. **No** canals were visible on any of the photographs. In fact, the space probes sent to Mars have caused us to reject much of what we felt to be fairly certain about the planet and have provided a virtual avalanche of information and discoveries. We will now turn our attention to these missions.

8.6 Early Space Probes to Mars

It always seems remarkable to me that so soon after the launching of *Sputnik 1* in 1957, which, after all, was the very beginning of what we may term the space age, space probes were successfully sent to the nearer planets. Not all the earliest space probes were entirely successful, though. The Russians were first to attempt to send a probe to Mars in 1962 when they dispatched *Mars 1* to the red planet, though contact with the vehicle was lost when it was only part of the way towards Mars.

The Americans were next with *Mariner 3*, launched in 1964, but it never actually arrived in the vicinity of Mars owing to a faulty takeoff. Then came *Mariner 4* in 1964, which flew by the planet in 1965, a mere 10 000 km above the planet and returned the twenty-one close-range pictures mentioned above. These showed that the dark regions of it were not as sharply defined as they had seemed from Earth-based observations. Also, the surface of the planet was not as flat as had been assumed. Indeed, at least one feature was seen that rose to a height of over 400 metres. Several craters were also observed, and the opinion was that Mars displayed a lunar-like landscape.

Altogether, *Mariner 4* had pictorially sampled a mere 1% of the surface area of the planet. Perhaps the most significant result from *Mariner 4* was the estimate of the atmospheric pressure on Mars. The spacecraft achieved this by sending radio signals back to the Earth as it was about to pass behind the planet. Just before the signals were cut off their distortion caused by the Martian atmosphere was analysed and this gave reliable information on composition and density. These results showed that the mean pressure at ground level was less than 1% of that of the Earth's atmosphere at sea level and that the main component of the Martian atmosphere was carbon dioxide.

The Russians again attempted to send a probe to Mars, *Zond 2*, in 1964 but, once again, contact was lost. In 1969 America sent *Mariners 6* and *7* (*Mariner 5* was a Venus probe) and each achieved a fly-by of the planet at a distance of a little over 3000 km, sending back many pictures. Many more craters were revealed, with diameters up to 200 km, and in the planet's winter the rims of many of these were covered in frosty deposits. Now opinions of Mars changed. Instead of being thought a planet where life was probable, if only in the vegetable form, it was now obviously a barren, desolate, almost lunar-like place of high mountains and large craters and with a poisonous atmosphere too thin to hold out any hopes of supporting life.

The space probes sent to Mars also studied the interplanetary medium and, in fact, detected no significant magnetic field or radiation zones surrounding the planet.

The Americans' next probe, *Mariner 8*, was a complete failure because it malfunctioned at launch and spectacularly crashed into the sea. The Russian probes *Mars 2* and *Mars 3* actually landed capsules on the planet, but no useful data was ever received from either craft. However, the glory of being first to actually land a man-made object on the Martian surface goes to the Russians. Later in the same year (1971) a big step forward in the study of the red planet came with the dispatch of the American vehicle *Mariner 9*.

8.7 Later Space Probes to Mars

It is a quirk of fortune that all the space probes until 1971 passed over the less interesting regions of Mars. Scientists jumped to the conclusion that all the planet was similar, but Mariner 9 was soon to change that view.

Mariner 9 was put into orbit just over 1600 km above the surface Mars, and sent back over 7000 pictures. These revealed a highly eroded surface; great valleys and canyons were discovered, many of which appeared to have been formed by the cutting action of a flowing liquid – perhaps water.

Several enormous volcanoes were also discovered. These are of the "shield" type and are the largest such structures in the Solar System. Moreover, Mars was found to be split into two distinct hemispheres. The southern hemisphere is more densely cratered and its landscape looks more like the Moon's highland regions. The northern hemisphere consists mostly of sparsely cratered plains. We do not know the reason for the difference between the two hemispheres.

Nearly all the young volcanic features are found in the sparsely cratered northern hemisphere. Volcanoes have been found in the southern hemisphere as well, but these all seem to be much older and highly eroded. The largest volcano on Mars, Olympus Mons (previously seen from Earth as a light spot – the Nix Olympica), is an incredible size. It has a shield over 500 km across and a summit height of 23 km above the surrounding plains. Figure 8.4 (*overleaf*) shows a later space probe's aerial photograph of this amazing feature.

To say that the *Mariner 9* results caused surprise in the astronomical community would be an understatement. Probably the most exciting discovery was of the features that so closely resembled dried river beds. If they were really formed by running water, where is this water now? Certainly the present atmospheric pressure on Mars is too low to allow for the presence of liquid water (it would vaporise immediately). If once, long ago, running water did flow on Mars then

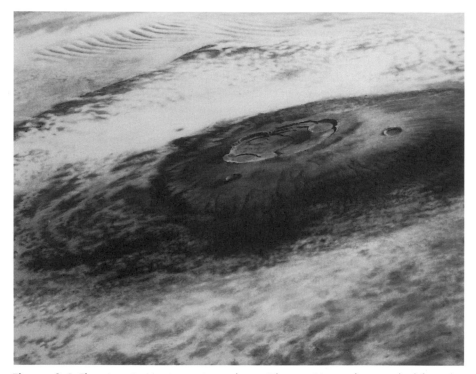

Figure 8.4 The gigantic Martian extinct volcano Olympus Mons, photographed from the *Viking* orbiter, is the largest volcano in the Solar System. Photograph courtesy JPL.

the atmospheric pressure must have been higher in those past times. Might the conditions then have been suitable for life to get a foothold?

Unfortunately, the spacecraft that followed *Mariner 9*, the Russian series *Mars 4* to *Mars 7*, were failures and we had to wait until 1975 for the next major advance – that of the two *Viking* probes.

8.8 *Vikings* To Mars

Two of the most ambitious probes ever to leave the Earth were launched in the late summer of 1975 – *Vikings 1* and *2*. They were each two-part vehicles, one part of which would orbit Mars while a specially designed lander descended to the surface in order to take various scientific measurements and even to take and analyse soil samples for signs of life!

After a revision of the selected landing sites, on 20 August 1976 the lander of *Viking 1* came down in Chryse (in the northern hemisphere). The probe immediately began transmitting pictures (see Fig. 8.5), many of which were in colour. These pictures showed a landscape of sand-dunes and pebbles, all of a rusty-red colour, which was attributed to the mineral limonite, which is mostly hydrated iron(III) oxide.

Even the sky was discovered to be a pinkish colour and it is thought that this is due to fine particles of dust swept from the surface of the planet by the winds. The atmosphere was found to be composed of 95% carbon dioxide, 2.7% nitrogen, 1.6% argon and 0.15% oxygen, the remainder being trace elements.

Viking 2, which landed on 3 September 1976 in Utopia, found a rather different landscape (Fig. 8.6, *overleaf*). Here the area round the lander was cluttered with rocks of varying types – some fine-grained and some vesicular (porous and sponge-like), with small pebbles and fine sand in between the larger rocks.

While most of the attention of the press focused on the pictures sent back by the landers, the orbiters were equally busy sending back much useful data and many spectacular pictures. The great canyons, cliffs, craters and volcanoes (all inactive) and ancient river-beds of Mars were seen in fabulous detail (see Fig. 8.7, *overleaf*).

Temperatures over the planet were measured. At local noon on the equator the temperature rose to a pleasant 15 °C, and at night fell to about –80 °C. Perhaps one of the most significant results was that the temperature of the northern polar cap was measured at about –70 °C. At the prevailing atmospheric pressure on Mars this is above the temperature at which carbon dioxide could persist in a frozen form. Together with spectroscopic evidence, the view is now that the polar caps are mainly composed of water–ice.

The polar caps of Mars are thought to be about 1 km thick. Instead of Mars being virtually waterless, we now realise that water is relatively plentiful. The poles contain large volumes of frozen water, and much is bound up in chemical combination in the soil. Also, the *Viking* orbiters have found substantial amounts of water vapour in the atmosphere.

Figure 8.5 The Chryse Planitia area of Mars as seen from the lander craft of *Viking 1*. Courtesy JPL.

Figure 8.6 Utopian Plain on Mars, as seen from *Viking 2*. A small part of the lander itself can be seen in the lower left of this photograph. Courtesy JPL.

It now seems that the polar caps of Mars are composed of two parts – a permanent cap made up of frozen water and a seasonal covering of carbon dioxide frost. The relative abundance of water once again raised hopes of finding some evidence of Martian life.

8.9 Life on Mars?

The famous astronomer Sir William Herschel thought that life on Mars was an absolute certainty – but then he thought that virtually every body in the Solar System was inhabited by thinking creatures, including the Sun! Percival Lowell was a great believer in intelligent life on the planet and cited the "canals" as strong evidence. The views of these two astronomers were expressed at the ends of the eighteenth and nineteenth centuries, respectively.

By the mid-twentieth century the presence of intelligent life, or any form of biologically advanced life, was strongly doubted. *Mariner 9* caused great pessimism among those who had hoped to find some form of life on Mars, if only the vegetable life whose presence had been confidently expected before then. Would the *Viking* craft supply a conclusive answer to the question of life on the red planet? Many thought so.

The Viking landers were each equipped with a scoop to dig up soil samples, which were analysed and subjected to various experiments on board the craft.

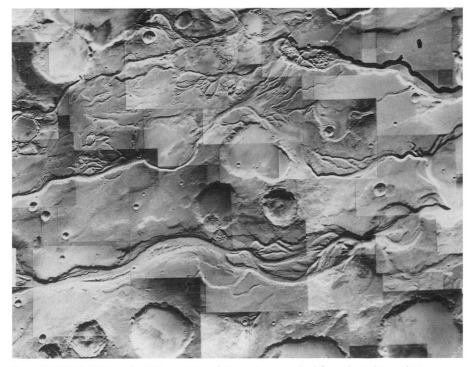

Figure 8.7 The Mangala Valles region of Mars photographed from the *Viking* orbiter space-craft. Courtesy JPL

Samples were heated to see if any ingestion of carbon-14 (from a supply of carbon-14-"labelled" carbon dioxide gas) took place. No conclusive result was obtained from either lander.

Next, carbon-14-labelled nutrients were added to samples of the Martian soil in order to detect possible microbial activity–again the results were neither entirely negative nor entirely positive.

To other samples water, and then water plus nutrients, were added, and any gaseous emissions were measured. At first there was a flurry of activity from the samples but after a short while they became inactive. It is now thought that the results obtained were **probably** due only to chemical action (involving superoxides) and so once again the possibility of current life on Mars, at least at the microbial level, is thrown open to debate.

However, as I am about to dispatch this manuscript to the publishers (August 1996) a news story has just broken that a NASA laboratory has found some evidence of possible past life on the red planet. Scientists have been examining a meteorite which has been thought to have been "chipped off" Mars about 16 million years ago when an asteroid impacted with it. This particular rock eventually arrived on Earth (eleven others have also been identified as being Martian as a result of chemical analysis). It seems that this rock contains some of the "chemical fingerprint" left by life processes, and even some microfossils left by tiny organisms have been reported.

If these reports and the assertions made from them are substantiated, then the consequences are immense. Life on another planet (even it was millions of years in the past) in our own Solar System surely must mean that the Universe is likely (though still not certain) to be teeming with life. I think that I am safe in predicting that the heavy financial cut-backs in space exploration that were being undertaken at the time of writing will be reversed as a consequense of this find. Whether or not the NASA scientists have made a mistake, either in their identification of the rock as coming from Mars, or in their investigations of it, should be clear by the time that this book appears in print.

8.10 Phobos and Deimos

Of the inner, rocky, planets of the Solar System, known collectively as the *terrestrial planets,* Mercury is known to have no satellites and neither has Venus. Of course, the Earth has one (the Moon) but until 1877 the two Moons of Mars had remained undiscovered.

In that year Asaph Hall, an astronomer in the USA, was occupied on a deliberate search for a satellite of Mars, using the 26 inch (660 mm) refractor at the Naval Observatory, Washington, which was then the largest refracting telescope in the world. At first he was unsuccessful but then, within a single week, Hall found the two satellites, which he subsequently named Phobos and Deimos (meaning "fear" and "flight").

The inner satellite, Phobos, orbits the planet at a mean distance of 3600 km with an orbital period of 7 hours 39 minutes and thus goes round the planet quicker than the planet rotates on its axis. Deimos orbits the planet at a mean distance of 9100 km, with an orbital period of 30 hours 18 minutes.

The *Mariner* and *Viking* space probes have found both satellites to be small, irregularly shaped and rocky. Phobos is about 9 km in diameter and Deimos is about 5 km in diameter, and both are very dark and cratered.

The satellites are difficult to detect from the Earth because they never appear far from the image of Mars, and are thus swamped in the bright glare from the planet. If the satellites could be seen on their own they would appear to be star-like points of the twelfth magnitude.

Questions

1   What is meant by the statement "Mars is a superior planet"? Describe, in detail, the orbit and phases of the planet Mars. Explain what is meant by *opposition* and explain why oppositions of Mars are not equally favourable to observers of the planet.

2   Describe the appearance of the planet Mars as seen through a moderately sized telescope (say a 200 mm reflector), explaining the nature of any observed features.

3 What are the "canals" of Mars? Give a brief history of their study. Why were the canals taken as evidence for intelligent life on the planet?

4 Describe the surface features and surface conditions (temperature, atmosphere, etc.) on Mars as we know it today.

5 Compare and contrast our knowledge of the planet Mars (a) before the space age and (b) today, emphasising where some of the older ideas were found to be wrong.

6 Describe as fully as you can the apparent path of Mars across the celestial sphere.

7 Give a **brief** history of the space probe investigations of Mars, specifying those that gave the most important results and stating what those results were. (It is **not** necessary to go into very detailed accounts of any of the probes, nor to give the failures more than a passing mention.)

8 Outline the discoveries of, and the main facts known about, the two natural satellites of Mars.

Chapter 9

The Planet Jupiter

THE inner, small, rocky worlds of the Solar System collectively form the *terrestrial planets*. Beyond these worlds lie the large gaseous bodies of the planets Jupiter, Saturn, Uranus and Neptune. These are collectively known as the *gas-giant planets*.

Jupiter is the largest planet in the Solar System, having a volume 1300 times that of the Earth. The mass of the planet is equivalent to 317 Earth-masses and so its average density is very much less than that of the Earth, being a mere 1300 kg/m^3. It orbits the Sun at a mean distance of 778 million kilometres, which is 5.2 astronomical units (AU). The orbit is reasonably circular.

9.1 The Telescopic Appearance of Jupiter

Oppositions of the planet occur at intervals of roughly 13 months and, except when Jupiter is near conjunction with the Sun, it is very easy to observe. This is particularly so because of its size. At opposition the planet subtends an apparent diameter of 50 arc seconds, shrinking to 30 arc seconds when near conjunction. At opposition, a magnifying power of only about ×40 will render the image of the planet through the telescope the same size as the full Moon seen with the unaided eye. To the naked eye Jupiter appears as a bright, slightly off-white "star". Jupiter is always more brilliant than Sirius and attains an apparent magnitude of –2.5, so that it really cannot be mistaken for anything else.

Even through a small telescope it can been seen that the disk of the planet is not perfectly round – Jupiter bulges at the equator at the expense of being flattened at the poles. The figures for the equatorial and polar diameters are 143 800 km and 133 500 km, respectively. This amounts to a flattening of one-fourteenth.

Small telescopes will show that Jupiter is crossed with light and dark bands, termed *zones* and *belts*, respectively. Large-aperture telescopes reveal many of these, together with finer detail. It is obvious from observing the planet that the visible surface is decidedly fluid in nature. Many of the details are transient and change their appearance. In addition, as the various features are carried across

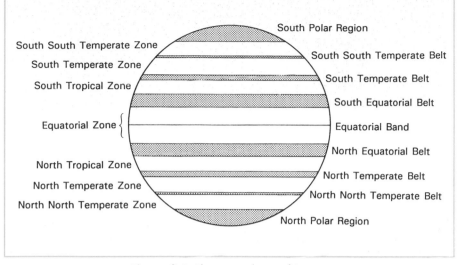

Figure 9.1 The nomenclature of Jupiter.

the disk by the rotation of the planet they travel at different rates. In fact, the apparent rotation of the planet as deduced from observations of features on the planet's equator is faster than that deduced from observations of features moving at higher latitudes.

It is obvious that our telescopic observations are of Jupiter's cloud-topped atmosphere, and any solid surface lies well below the levels observable from the Earth. Owing to its large apparent diameter, the physical study of the planet was well advanced by the time that spacecraft visited it in the 1970s.

Astronomers have standardised the nomenclature of the visible surface features, and the main scheme is given in Fig. 9.1. The belts near the equatorial regions of the planet are fairly permanent features, though they often display variations in width, colour and intensity. They also drift a little in latitude. The higher-latitude belts are of a less permanent nature, usually being much less prominent and often merging with the polar hoods. The zones are usually pure white to pale yellow in colour, whilst the belts are usually various shades of orange, red or brown. The average albedo of the visible disk of the planet is 0.51. Figure 9.2 (*overleaf*) shows an observation of the planet I made using my own telescope.

9.2 Transient Features

Projections and other irregularities are often seen in the belts and zones and dark coloured spots sometimes appear. These features may appear, develop, and then vanish in a few hours, a few days or a few months, and the telescopic observer nearly always has a variety of interesting detail to study. We have long known that all these features are purely atmospheric in origin, though we are still not sure of all the details of the mechanisms in operation.

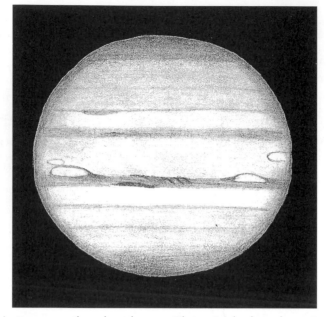

Figure 9.2 Jupiter as seen through a telescope. Observation by the author, using his $18\frac{1}{4}$ inch reflecting telescope, ×288, on 1977 December 30^{d} 23^{h} 25^{m} UT.

Jupiter's atmosphere moves in vast circulating currents. The spots appear to be caused by eddies in adjacent jet streams. Two major circulation rates exist between latitudes 60° north and south. The Equatorial Zone (10° either side of the equator) possesses an average rotation period of 9 hours 50 minutes 30 seconds, whilst the higher latitudes rotate with an average period of 9 hours 55 minutes 41 seconds. In order to set up a longitude system on the Jupiter, the equatorial and higher latitude areas of the planet are related to *System 1* and *System 2*, respectively. Each system is related to the region's average rotation rate. This allows the study of longitude drifts on a comparative basis.

In order to do this, the observer times the passage (or "transit") of a given feature across the centre line of the disk (running north to south). This line runs from pole to pole and is termed the *central meridian*. The longitude on the meridian at the time of observation is found from published tables, relating either to System 1 or System 2, depending on the observed latitude of the feature.

Although all the visible Jovian phenomena so far discussed are purely atmospheric features, they sometimes display remarkable longevity. This is particularly true of the *Great Red Spot*.

9.3 The Great Red Spot

The Great Red Spot (GRS) is an immense elliptical area situated in the South Tropical Zone which has been visible for at least 300 years. The first person definitely to record its appearance was the astronomer Cassini in 1665, although it is possible that an observation made by Robert Hooke in 1664 actually refers to the feature. Over a period of time the GRS exhibits changes in latitude, longitude, size and colour. At some times it is the darkest feature on the planet, being a strong reddish hue, while at other times it becomes virtually invisible but leaves a white "hollow" in its place. It was a memorable strong salmon pink in the mid-1970s but faded almost to invisibilty in the mid-1980s. The spot has darkened again since, but is still not as prominent as in the 1970s.

Historical records show that the GRS was at its largest in the 1880s, but it has since shrunk to a length of about 26 000 km. This is about half its former value. Its width has remained fairly constant at about 14 000 km, and so it is now very much less elliptical than it used to be. Spectral analysis from the Earth showed the presence of large amounts of hydrogen, methane and ammonia in the atmosphere of Jupiter, and infrared measures had indicated that the average cloud top temperature was in the region of –140 °C. It was known that the GRS shows up as a particularly cold spot. Many theories were concocted to explain the appearance and longevity of the feature, but observations from space probes were required before we really began to make sense of this peculiar phenomenon.

9.4 *Pioneers 10 and 11*

Pioneers 10 and *11* bypassed Jupiter in the Decembers of 1973 and 1974, respectively. A large number of coloured images were taken and radio-relayed back to Earth, as were other instrumental data. These showed that the spots, including the GRS, are great swirling vortices in the atmosphere. The zones and belts consist of gaseous substances and liquid and solid particles, at different altitudes and temperatures.

The zones are the highest and coldest of the atmospheric features. They consist mainly of frozen ammonia crystals and this explains their whiteness. At lower levels in the 1000-km-thick atmosphere the temperature is above the freezing point of ammonia. It is in these levels that most of the coloured compounds reside.

Infrared measures proved surprising – these indicate that Jupiter radiates twice as much heat into space as it absorbs from the Sun. The collected data have been used to construct theoretical models of Jupiter's atmosphere, and these models predict that the pressure and temperature increase rapidly with depth. It is the outflowing convective heat currents that produce the differential features. In other words, gas warmed by the planet's internal heat decreases in density and rises into the upper atmosphere (in accordance with Archimedes' law of flotation), forming clouds of ammonia crystals suspended in gaseous hydrogen. At the edges of the clouds, the ammonia descends into the lower, coloured, atmospheric layers.

The rapid rotation of Jupiter (which causes the appreciable flattening of the globe) generates coriolis forces which wrap the clouds around the planet, so producing the characteristic series of belts and zones. Dynamic instabilities result in the formation of the various transient loops and swirls, together with the spots. The colours are thought to result from free sulphur and phosphorus, as well as their compounds.

9.5 The *Voyager* Missions

We learned much from the two *Pioneer* probes but more was yet to come from the *Voyager* craft in 1979. American space scientists took advantage of a rare alignment of the outer planets to launch two probes that would use the gravity of each major planet to "sling-shot" the probes on to the next. The two probes, *Voyager 1* and *Voyager 2*, were launched in the summer of 1977. *Voyager 1* bypassed Jupiter in March 1979 and Saturn in November 1980. This probe then continued on its way, heading out of the Solar System and into interstellar space. *Voyager 2* passed Jupiter in July 1979, Saturn in 1981, Uranus in January 1986 and Neptune in September 1989, leaving Pluto as the only major planet not yet visited by a space probe.

The two probes are identical, each carrying ten scientific instrument packages, including television cameras. As *Voyager 1* approached Jupiter, it became obvious that the atmospheric motions were more complex than scientists had first thought. Small-scale vortex motions dominated the planet-wide picture. The counter-flowing atmospheric currents and jet streams were studied in remarkable detail and a large number of cloud and spot interactions were studied (see Fig. 9.3).

As a result of the two *Voyager* probes, we now think that features like the Great Red Spot are *solitons*. These are solitary waves that develop as vortices between currents that flow with different velocities. The GRS appears to be a vast anti-cyclonic (anticlockwise) whirlwind which draws material up from the lower levels of the atmosphere to well above the clouds of frozen ammonia. The *Voyager* craft were highly successful, leading to a significant increase in our knowledge of the outer planets and their satellites.

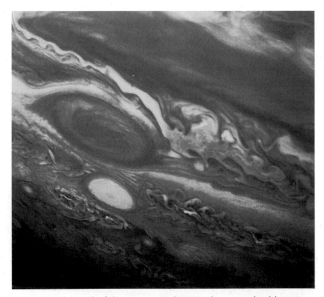

Figure 9.3 The neighbourhood of the Great Red Spot photographed by *Voyager 2*. Courtesy JPL

9.6 Collision with a Comet

Perhaps one of the most remarkable events ever to be seen by astronomers was the impact of the fragments of the disintigrated comet Shoemaker–Levy 9 on the planet Jupiter in July 1994.

There is more about comets in Chapter 13, but suffice it to say here that a comet is mainly an extremely tenuous mixture of gas and dust. However, the source of this gas and dust is a small (a couple of kilometres across, or so) "dirty snowball" of various chemical ices mixed with fine rocky fragments. This *nucleus* is the only really substantial part of a comet.

Eugene and Caroline Shoemaker and David Levy were conducting a photographic search for new asteroids and comets using the 18-inch (46 cm) Schmidt camera at Mount Palomar when they came across the fateful comet in March 1993. Immediately it was apparent from the photographic plate that something was odd about this object. Other telescopes were turned towards it and it was confirmed that the comet's nucleus had fragmented. Later analysis suggested that it had been tidally disrupted on its previous very close orbital fly-by of the planet Jupiter in July 1992. Astronomers soon realised that this time round it would not merely fly by Jupiter. It was going to collide with the planet.

The excitement of astronomers mounted as they watched the separate nuclei, like a string of cosmic pearls, head towards the great planet. Some of the fragments were very small. Indeed, some vanished altogether, seemingly evaporated out of existence, while others were more substantial. Some of the surviving fragments further subdivided, indicating that they were rather flimsy in nature. As the predicted date of the first impact approached more and more telescopes were turned towards Jupiter, and the story made international news.

Then on 16 July 1994 at $20^h 12^m$ UT the first of about twenty-one separate fragments impacted with the Jovian upper cloud decks. The bombardment continued a further six days. Although almost all of the impacts occurred just round the limb of the planet, and so out of sight of astronomers, infrared telescopes (such as UKIRT on Hawaii) typically saw the flash of each fragment and its surrounding gaseous envelope ploughing into Jupiter's atmosphere, followed by a great fireball and plume of atmospheric gases erupting following the vast explosion.

As each impact site rotated into view astronomers were shocked to see enormous black "scars" on Jupiter's cloudy mantle. At that time Jupiter was only visible very low down in the twilight sky from the UK. However, the impact scars were extremely easy to see even with a small telescope, despite the poor observing conditions, as Fig. 9.4 (*overleaf*) shows. Most of the impact scars lasted more than a month, changing shape and structure all the time, and Jupiter eventually developed a new, temporary, cloud belt as the surviving impact scars became stretched around the globe; by 1996 the planet had returned to normal. However, though the ill-fated comet Shoemaker–Levy 9 is gone, it will certainly not be forgotten. This spectacle provided astronomers with the opportunity of studying both a comet and Jupiter in a way never before invisaged, and much complex physics and chemistry was learned about them as a result.

Figure 9.4 The aftermath of the impacts of the comet Shoemaker-Levy 9 on Jupiter can be seen in the observation made by the author, through his $6\frac{1}{4}$ inch (158 mm) Newtonian reflector, ×203, on 1994 July 20^d 20^h 21^m UT.

9.7 *Galileo* Enters the Jovian Atmosphere

After a decade of delays before its launch and a six year journey, the two-part *Galileo* spacecraft finally arrived at Jupiter in December 1995. One part, the probe, successfully parachuted into the Jovian atmosphere. The other part, the Orbiter, set itself into an orbit about the Jovian planet that took it through the Jovian satellite system. Various malfunctions had also troubled the spacecraft on its way to Jupiter, and much of what was planned had to be scrapped. However, some of the problems were overcome and some images (from the orbiter only) and useful data have been received. The mission was expected to continue to about the end of 1997.

It was the information sent back by the probe that proved particularly interesting to planetologists, as this was the first time any such active device had penetrated below the clouds of one of the gas-giant planets. The initial results are interesting and, as usual, some of them are unexpected.

The probe registered an increase in pressure from 0.3 atmospheres (1 atmosphere = Earth's atmospheric pressure at sea level) near the cloud tops to 22 atmospheres, a distance of roughly 160 kilometres below that point. The corresponding atmospheric temperatures were –144 °C increasing to 152 °C, at which point the probe ceased to function, 57 minutes after it had begun to transmit. The decending probe was swept along by winds of around 150 m/s. These were expected to lessen with depth but, if anything, the reverse seems to be the case.

The first indications are that the expected upper cloud layer of ammonia crystals was registered. Below this it was expected to find clouds of ammonium hydrosulphide, and at the time of writing it seems that these have also been found. Below that, however, they had expected to find clouds of water crystals, but to everybody's surprise the atmosphere seemed to be incredibly dry, confounding the meteorologists models.

Scientists also expected the probe to register the characteristic radio signature of lightning. It didn't.

There are possible explanations for these unexpected results, and it should be emphasised that the probe has sampled only one very small location on the vast planet (it dropped through a clearing at the edge of the North Equatorial Belt), so the data it obtained might be atypical. Anyway, the analysis is still far from complete, though a clearer idea may emerge quite soon.

9.8 The Structure of Jupiter

The *Voyager* results confirmed the belief that the composition of Jupiter is very similar to that of the Sun – roughly 90% hydrogen and 10% helium, with very minor amounts of other elements and compounds. Using this as a basis, scientists have agreed on the probable structure of the planet.

It is thought that the gaseous envelope of hydrogen and hydrogen compounds extends down to a depth of about 1000 km below the cloud tops. Both the pressure and the temperature increase with depth. At this level is a planet-wide ocean of liquid hydrogen. The pressure at this depth would be about 5600 atmospheres and the temperature would be in the region of 1700 °C. At a depth of about 25 000 km, where the pressure and temperature have increased to 3 million atmospheres and 12 000 °C, the liquid molecular hydrogen then reverts to its liquid metallic state. In this form, hydrogen behaves rather like a metal in that it can conduct electricity.

We think that there is a rocky core at the centre of Jupiter, with a mass perhaps fifteen times that of the Earth. The metallic hydrogen zone is thought to extend down to this core, where the temperature reaches about 30 000 °C and the pressure reaches the huge value of 100 million atmospheres.

The acceleration due to gravity at the visible surface of Jupiter is 2.6 times that at the surface of the Earth. This has prevented much of Jupiter's primordial atmosphere from leaking away into space, and so the composition of the planet has not radically changed since its formation, 4600 million years ago. The composition of Jupiter leads us to regard this planet as a "failed star", a body not quite massive enough to enable its central temperature to reach the value where nuclear reactions can begin. It is the nuclear processes going on within its core that have kept the Sun shining all these aeons. Nevertheless, the heat that Jupiter pours out into space is thought to be derived from the gradual contraction of the planet.

9.9 Jupiter's Magnetosphere

As might be expected from its metallic hydrogen mantle and its rapid rate of rotation, Jupiter has a magnetic field of a similar overall structure to the Earth's field but on a far grander scale. The magnetic field strength at Jupiter's cloud tops is 12 times that at the surface of the Earth. The vast magnetosphere extends from 25 to 50 times the diameter of Jupiter in the Sunward direction, whilst the magnetotail extends beyond the orbit of Saturn. As is the case with the Earth's magnetosphere, the gusty solar wind causes variations in the shape and size of Jupiter's magnetic field, though the variations are greater in the case of Jupiter.

9.10 Radiation Belts

In the same manner as the Earth's magnetic field, the Jovian field traps particles to form enormous radiation zones around the planet. In the case of Jupiter, the zones extend out to about ten times the diameter of the planet. The maximum radiation intensity is 10 000 times stronger than the peak intensity within the Earth's Van Allen zones. *Pioneer 10* came close to being put out of action by the intense radiation, and *Pioneer 11* was rerouted to carry it over the south pole of Jupiter and so rapidly pass through the most dangerous regions.

9.11 Radio Emissions

It was a major surprise when, in 1965, intense and fluctuating radio emissions were detected from Jupiter. This was, correctly, taken to indicate that Jupiter possesses a magnetosphere. The radio emissions could only be explained in terms of charged particles interacting with a magnetic field.

Apart from the Sun, the planet Jupiter is the strongest emitter of radio waves in the Solar System. The emissions can be divided into three distinct components, each categorised in terms of wavelength. The first is the decametric range, having wavelengths of from 7 to 700 metres, and consisting of irregular and sporadic pulses of "radio noise". Soon after their discovery, it was realised that these pulses are related to the apparent positions of the moon Io, but the mechanism is still not well understood.

The second category is the decimetric range, of wavelengths downward from 7 metres, consisting of a steady stream of "radio noise". We now know that this radio emission is produced by electrons spiralling along the Jovian magnetic field lines. These electrons lose energy by a process known as *synchrotron emission*. It is interesting to note that the polarisation of the decimetric radio emission indicates that the polarity of Jupiter's magnetic field is opposite to that of the Earth.

The third catagory is thermal emission. This is also a steady emission, being due to the temperature of Jupiter's outer layers. The thermal radiation actually peaks in the infrared part of the spectrum (the wavelength being about 25 μm) but tails into the microwave and radiowave portions.

9.12 The Satellites

Use a small telescope to observe Jupiter over a period of a few nights and it will be obvious that at least four moons, or satellites, orbit the planet. These four were independently discovered by Galileo and Simon Marius at about the same time in 1610. They have become known as the *Galilean satellites* and have the names given to them by Marius: Ganymede, Callisto, Io and Europa. These objects are bright and are easily seen through a good pair of binoculars.

On most nights the Galilean satellites can be seen strung out in a line passing through the equator of Jupiter and extending out to several Jupiter-diameters from it. They all orbit at different distances from their parent planet and so move round with different orbital periods. Hence, the satellites are constantly changing their configurations. At various times they each pass behind the disk of Jupiter, as seen from the Earth. They are then said to be *occulted* by the planet (and sometimes its shadow, as this can extend slightly to one or other side of the planet as we see it from Earth). The satellites are said to *transit* Jupiter when they pass in front of its disk.

The satellites can even, though rarely, occult and transit one another. Telescopes of around 12 inches (30 cm) aperture can show the Galilean satellites as definite disks and a telescope even smaller than this will show the shadows cast by them on Jupiter's cloud tops. A fifth satellite, named Amalthea, was discovered in 1892. It is very much fainter than the other four (magnitude 14, whereas the Galilean moons range from fourth to fifth magnitude). Many more faint satellites have been discovered this century, bringing the total up to sixteen.

The satellites can be divided into three groups, according to their orbital properties. Closest to the planet there is a group of eight satellites that include the Galileans. They all have fairly circular orbits, which lie in the same plane as the planet's equator. This group ranges from the small body discovered by the *Voyager* research scientists in 1980, orbiting at a mean distance of 128 000 km from Jupiter, to Callisto, orbiting at nearly 2 million km from the planet.

The second group consists of four bodies, all orbiting at a mean distance of about 12 million km. Their orbits are much more eccentric than the inner satellites, as well as being steeply inclined to Jupiter's equatorial plane. The third group of satellites, possessing four members, orbit Jupiter at mean distances of 21 to 25 million km. They also have highly eccentric and steeply inclined orbits but are rather odd in that they orbit Jupiter the wrong way, or in the retrograde direction. Astronomers think that the four outer members of Jupiter's satellite family did not form with the planet but were later captured when they wandered into its gravitational influence.

Most of Jupiter's moon's are rather small, insignificant, bodies with the exception of Io, Europa, Ganymede and Callisto. Indeed, the smallest of these, Europa, is only slightly smaller than our Moon. The largest, Ganymede, is a little bigger than the planet Mercury.

9.13 Io

Io orbits Jupiter at a mean distance of 422 000 km, with a period of $1\frac{3}{4}$ Earth-days. It has a diameter of 3600 km and a density of 3500 kg/m^3. This makes it the most dense of the Galilean satellites. From the Earth, Io looks distinctly yellowish but little in the way of surface detail can be seen. All the Galilean satellites subtend an angular diameter of little more than 1 arc second as seen from the Earth.

Voyager 1 revealed Io to be a spectacular world (see Fig. 9.5). Most of the published pictures of the planet show the overall colour of the surface as orange with patches of various shades of red, yellow and black as well as glistening white deposits. However, these are enhanced colours and the true tones are somewhat more muted. The surface is mainly covered in sulphur, traces of the various allotropes of this element producing the different shades. Erupting volcanoes were in evidence at the time of the Voyager mission, and have been detected from the Earth by their infrared signatures since. It is clear that the surface of the satellite is very young. Enormous plumes of sulphur spray from the erupting volcanic vents and continually modify the landscape.

The Galileo orbiter has found that Io has a magnetic field and it is thought to have a large, iron-rich, core. Planetary scientists are of the opinion that the tidal forces set up by Jupiter and the other satellites cause frictional heating in this core. This heat causes the volcanism. The white surface deposits are thought to be frozen sulphur dioxide, and a cloud of sulphur vapour has been detected about Jupiter, concentrated in a toroidal belt which marks the orbit of Io. Some scientists think that bombardment from particles in the radiation zones (Io moves through one of the more intense zones) causes some of the surface deposits to be splattered into space. Sodium has also been detected in the plasma torus.

Io has no appreciable atmosphere. With its erupting sulphur volcanoes, Io is like something out of science fiction. Certainly, the radiation bombardment must make it one of the most lethal places in the Solar System!

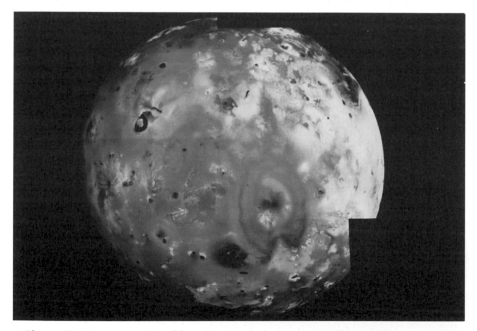

Figure 9.5 Io, a composite of four photographs by *Voyager 1*. Photograph courtesy JPL.

9.14 Europa

With a diameter of 3100 km, Europa is the smallest of the Galilean satellites. It orbits Jupiter at a mean distance of 671 000 km. The mean density of this satellite is 3300 kg/m^3 and it is thought to have a silicate core. The albedo of Europa is 0.64, slightly greater than that of Io (0.63) and it has a whitish surface, thought to be icy in nature. Europa has an incredibly smooth surface with very little vertical relief. It is also virtually devoid of craters and these factors have caused much debate amongst planetologists. The surface of the satellite seems inactive, so where are the craters one would expect on any ancient surface in the Solar System? By comparison, Ganymede and Callisto are smothered in craters.

Clearly, Europa's covering of "dirty ice" remained liquid until after the main cratering processes had died down in the outer Solar System. Even more puzzling is the fact that Europa is seen to be covered in a network pattern of thousands of dark lines. Perhaps, these lines represent fractures in the surface where "dirtier ice" has flowed up from below to fill in cracks in the surface caused by an expansion of the core of the planet? At present, we must admit that the apparent quiescence of Europa belies a little world with its own share of mysteries.

9.15 Ganymede

Ganymede is the largest of the Jovian system of satellites and is the largest satellite in the Solar System, having a diameter of 5270 km. It orbits Jupiter at a mean distance of 1 million km. Despite its low surface albedo (0.4) it is the brightest of Jupiter's retinue, as seen from the Earth. This satellite is much less dense than Io or Europa, the mean density being 1900 kg/m^3. Like Europa, Ganymede is basically icy with darker contaminants. Astronomers think that it has a large silicate core, overlaid by a slushy mantle of ice, in turn overlaid by a crust of about 100 km thickness. However, it must be admitted that our ideas of the interiors of all these bodies are tentative, to say the least.

As has already been mentioned, Ganymede is extensively cratered and the surface of the satellite bears a superficial resemblance to the surface of our Moon, with many bright craters and "rays". However, there are many differences. On Ganymede there are large plains of strangely lined terrain, and broad bands abound over the surface. Lateral faulting is evident and it seems that in past times an icy form of "plate tectonics" took place on this frozen globe.

9.16 Callisto

Beyond the orbit of Ganymede lies that of Callisto, 1.9 million km from Jupiter. Callisto is the least dense of all the Jovian satellites, with a value of 1600 kg/m^3. It is only slightly smaller than Ganymede, with a diameter of about 5000 km. Like Ganymede, Callisto is thought to be composed of ices, in the main, with a smaller silicate core. The albedo of Callisto is only 0.2, so it appears much fainter than the other Galilean satellites when viewed from the Earth, despite its size.

The surface covering on Callisto is thought to be "dirty ice" and is undoubtably very old. Indeed, its crater-saturated surface is thought to be the most ancient yet studied in the Solar System. As with Ganymede and Europa, there is little in the way of surface relief on Callisto. This is consistent with an icy surface. One striking feature is the presence of large systems of concentric rings, thought to be the frozen shock waves of ancient impact events.

9.17 The Rings of Jupiter

One of the most unexpected discoveries arising from the Voyager missions was that of a faint and tenuous ring system surrounding Jupiter. The outer edge of the ring, where it is brightest, extends to about 126 000 km from the centre of Jupiter. This is roughly 1.6 times the radius of the planet. The ring seems to be composed of fine rocky particles. Each of the particles orbits the planet separately, but all are in the equatorial plane of the planet. These give the appearance of a flat, if rather transparent, sheet when seen from a distance. The origin of planetary ring systems is discussed in the next chapter.

Questions

1   "Jupiter is a gas-giant world, while the Earth is a terrestrial planet." Explain this statement and briefly outline the main differences between the two worlds.

2   Describe the appearance of the planet Jupiter as seen through a small telescope and very briefly explain the natures of the visible features.

3   Describe the main facts learnt about Jupiter from the space probe investigations of the planet.

4   Write a short essay about Jupiter's Great Red Spot.

5   Write a short account of Jupiter's magnetosphere and radiation zones.

6   Describe the various phenomena of Jupiter's four chief satellites as they orbit the planet. Your answers should contain the terms *transit* and *occult*, explaining what they mean.

7   Describe the natures of the four Galilean satellites of Jupiter.

8   Write a brief account of the satellite family of Jupiter.

Chapter 10

The Planet Saturn

BEYOND the mighty Jupiter lies the second biggest planet of the Solar System – Saturn. Also a gas-giant world, Saturn has an equatorial diameter of 120 000 km and a mass 95 times that of the Earth. The average density of Saturn is a mere 700 kg/m³, less dense than water. Saturn moves through the Solar System at a mean distance of 1400 million km from the Sun, taking $29\frac{1}{2}$ Earth-years to complete one orbit.

10.1 The Telescopic Appearance of Saturn

Since Saturn moves relatively slowly about the Sun, oppositions of the planet occur only about a fortnight later each year. The planet then attains a maximum apparent magnitude of –0.4, appearing to the naked eye like a dull yellowish star.

Seen through the telescope Saturn presents a small, pale yellow disk, of 20 arc seconds apparent diameter at the time of opposition. The body of the planet is crossed by belts and zones reminiscent of those on the Jupiter, though all such features are less sharply contrasted than Jupiter's. Still, the same nomenclature is used to describe the belts and zones of Saturn, though only the Equatorial Belts and the Temperate Belts can be said to be permanent features of the planet.

If the main body of the planet appears only a smaller and more pallid version of Jupiter, this is more than made up by Saturn's impressive system of rings. Galileo was the first to observe the rings, though his telescopes were not powerful enough to show them in their true guise, and it was Christian Huygens, in 1659, who discerned, as he himself wrote: "a flat ring inclined to the ecliptic which nowhere touches the planet".

Soon it became evident that the rings were divided up into concentric segments (see Figure 10.1, *overleaf*). Giovanni Cassini, in 1675, discovered a line of darkness that separates the inner and outer components. This line has become known as Cassini's Division in his honour. The outer ring component, known as ring A, appears rather less bright than the inner ring, ring B. The rings are also different in colour – ring A is a slightly bluish grey and ring B is creamy white. In 1850

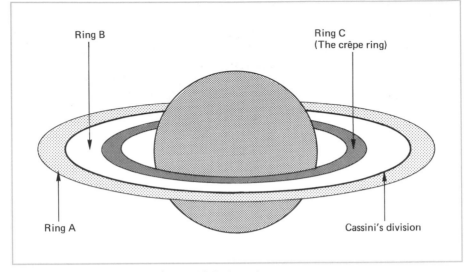

Figure 10.1 Saturn's main rings.

another ring component, ring C, was discovered by the British clergyman and amateur astronomer William Dawes and by George Bond in America. Both observers worked independently. Ring C is inside ring B and is not easy to see in telescopes smaller than about 6 inches (15 cm) in aperture, against the dark sky, because of the overpowering light from Saturn itself and its major rings. However, it is easily visible as a dark band where it crosses the globe of the planet, even through very small telescopes.

Apart from Cassini's, other divisions in Saturn's rings were reported from time to time, but only Encke's Division in the outer part of ring A was taken at all seriously until the space probes of recent years. Saturn's ring system is vast, spanning a diameter of over 270 000 km to the outer edge of ring A. The rings do not always present the same aspect to us, but rather yaw and tilt over a thirty-year cycle. There is no mystery about this, and the explanation is simply that our vantage point changes with respect to the planet as Saturn and the Earth move around the Sun. Reference to Fig. 10.2 should help to clarify the situation.

Saturn's rotation axis is tilted to the normal of its orbital plane by an angle of slightly less than 27°. Saturn, acting rather like a gyroscope, maintains the direction of its spin axis in space as it revolves about the Sun. Referring to Figure 10.2, when Saturn is at position S_1 it is at opposition when the Earth is at E_1 and so on.

In 1966 the rings of Saturn were presented edgewise to us and they were invisible as seen through small telescopes. Even through large telescopes they appeared no more than a thin line of light. By 1974 the ring system had opened out, so that the south face of the rings was on view and they were opened to their maximum extent. The rings then began to close up, becoming edgewise again in 1980. Continuing the cycle, the rings then opened, presenting their north face to us and becoming fully open in 1987, only to become edgewise once again in 1995. Saturn's ring system very accurately follows the equatorial plane of the planet.

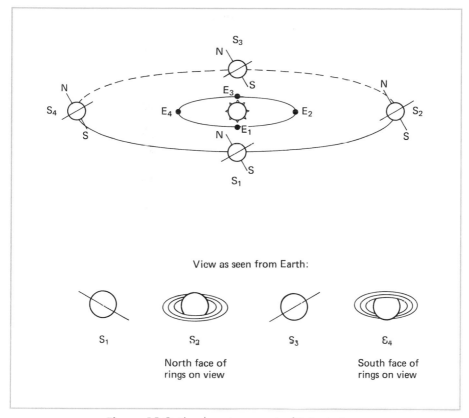

Figure 10.2 The changing aspects of Saturn's rings.

It has long been known that Saturn is a similar type of body to Jupiter, although the cloudy atmosphere appears more quiescent. Very occasionally white spots are seen and large telescopes do reveal irregularities in the belts; for example, the North Equatorial belt split into two components in the 1970s. Visible cloud-top features, when they occur, have been used to determine the rotation period of the planet. At the equatorial regions Saturn's clouds rotate with a period of 10 hours 14 minutes. It is this rapid rate of rotation, coupled with the fluid nature of Saturn, which leads to it presenting a disk even more flattened than Jupiter's. The polar diameter of Saturn is 107 000 km, as opposed to its 120 000 km equatorial diameter.

10.2 Spacecraft to Saturn

After successfully flying past Jupiter late in 1974, the space probe *Pioneer 11* was swung round by the Jovian gravitational field and sent out into space on a course for the planet Saturn, which it passed in September 1979. This probe contributed many interesting results on the magnetic environment about Saturn and the thermal properties of Saturn itself, as well as one of its satellites, Titan. However

Figure 10.3 Saturn as seen through a telescope. Observation made by the author, using his $18\frac{1}{4}$ inch reflecting telescope, ×260, on 1975 March 7^d 22^h 00^m UT. Note the shadow of the globe of the planet cast onto the rings and Cassini's division separating ring A from ring B.

the imaging results were disappointing. Because as little detail was seen in the Saturnian cloud belts and Titan looked even more bland.

The ring system provided the most interest; the probe was even scheduled to pass through the ring plane (to the trepidation of many project scientists) and passed through unscathed! Encke's Division was clearly photographed for the first time (no Earth-based photographs were well enough resolved to show it) and an additional ringlet was discovered.

Scientists were completely surprised by the spectacular discoveries made when *Voyager 1* bypassed Saturn in November 1980. These were many and varied and so it is best for us to consider each aspect of the planet separately.

10.3 The Atmosphere of Saturn

Spectroscopic analysis of the atmospheres of Jupiter and Saturn, carried out in the 1930s, had indicated that the compositions of the atmospheres of the planets were very similar, namely, hydrogen and hydrogen compounds, such as methane and ammonia. More recent infrared measures indicate that the temperature in the upper levels of Saturn's atmosphere is about –180 °C. This is cooler than at Jupiter's cloud tops, but nevertheless warmer than was to be expected if the Sun is the only source of the planet's heat. It seems that Saturn, like Jupiter, radiates more than twice the heat it receives from the Sun.

A haze layer above the clouds is partly responsible for the muted contrasts of Saturn's visible features, though we are not sure of its composition. The computer-enhanced images from the *Voyager* craft (*Voyager 2* followed *Voyager 1* to Saturn, bypassing the planet in August 1981) revealed a cloud morphology outwardly similar to that of Jupiter.

Puffy and chevron-shaped clouds were found, as well as a multitude of white spots and even a small red spot. There are differences between the atmospheres of Jupiter and Saturn, particularly in the distribution of the various wind currents. For instance, a wide band of cloud centred on Saturn's equatorial regions is moving at remarkabble 1800 m/s to the "internal" relative rotation rate of Saturn (as determined by radio observations of the emissions of Saturn's interior).

Saturn's atmosphere seems more thoroughly mixed, in terms of the different chemical constituents, than is the case with Jupiter's atmosphere. This, together with the haze layer, partly explains the absence of the bright colour contrasts so characteristic of the Jovian envelope.

10.4 The Structure of Saturn

Like Jupiter and the Sun, Saturn is essentially composed of about 90% hydrogen and 10% helium. We think that it possesses a 30 000 km diameter core, made of silicates and ices, overlaid by a 1500-km-deep mantle of metallic hydrogen, over which is a planet-wide ocean of liquid hydrogen nearly 30 000 km deep. Over this lies the cloudy atmosphere. The pressure at the surface of the core is thought to be about 10 million atmospheres, while the temperature there is 14 000 °C.

According to the theorists, gravitational contraction is only responsible for part of the heat energy that Saturn releases into space. They think that the conditions are correct for droplets of helium to form in the upper part of the metallic hydrogen layer. These droplets then make their way down towards the centre of the planet, releasing gravitational potential energy as they do so. In other words, a sort of "internal rain" of helium is responsible for most of the observed release of the planet's internal heat energy.

A byproduct of this proposed mechanism is that it neatly explains an observed depletion of helium in Saturn's outer layers. The outermost regions of the planet appear to be composed of about 93%, hydrogen and about 7% helium, but the average composition of the globe is thought to resemble that of Jupiter, as indicated earlier.

10.5 Saturn's Magnetosphere

Saturn's magnetosphere, though large by planetary standards, is much smaller and weaker than Jupiter's intense field. A measure of the overall strength of the magnetic field is given by a quantity known as the *dipole moment*. For Saturn this is 550 times that of the Earth's field. Jupiter's field has a dipole moment ten times stronger still. However, the magnetic field strength (also known as the *flux density*) at Saturn's cloud tops is roughly equal to the magnetic field strength at the surface of the Earth. The value of the dipole moment takes into account the physical size of the magnetic field as well as its strength. Saturn's magnetosphere extends to about ten Saturn-diameters in the Sunward direction. A long magneto-tail extends in the opposite direction, as is the case for Jupiter. Like Jupiter's magnetic field, north and south are in the opposite direction to the Earth's field. Saturn's magnetic field is unique in that the magnetic and rotation axes are aligned to within one degree.

A peculiar radio emission was detected by the *Voyager* craft on their approach to Saturn. The emission appears to emanate from two sources, one in the northern hemisphere of the planet and one in the southern hemisphere. Neither source rotates with the planet, but both remain aligned to the Sun, though the strength of the radio emission varies with a period of 10 hours $39\frac{1}{2}$ minutes. Scientists guess that this period reflects the rotation of the inner portions of Saturn's globe, though nobody yet understands the mechanism that produces the radio emissions. Radiation zones surround Saturn, though these are much less intense than those around Jupiter.

10.6 The Rings of Saturn

Saturn's ring system is a magnificent spectacle even in a small telescope, and it has long been realised that it cannot be made of a solid sheet of matter. In accordance with Kepler's laws, the inner parts of the ring system would "want" to move round the planet rapidly, while the outer parts would "want" to move more slowly. If the rings did rotate as a solid sheet then the outer parts would be forced to move more rapidly than the inner zones. The result is that the rings would be torn apart by shearing forces.

In 1895 James Keeler turned a spectroscope on Saturn's ring system and measured the Doppler shift of light (see Chapter 15) from various parts of it. In doing so he proved that the rings were indeed composed of innumerable particles each orbiting the planet and obeying Kepler's laws.

Optical studies of Saturn's rings have shown us that most of the particles composing them range in size from a centimetre, or so, to several metres. The particles are very reflective and are thought to be composed of ices, though there are differences in colour and reflectivity between the different ring components.

The Saturnian ring system is incredibly thin and flat. Before the space probe observations we knew that the rings could be no more than 18 km thick. In fact, they have a thickness of less than 1 km. This is rather remarkable, considering that the bright rings span over 280 000 km. From time to time other ring components and divisions in the rings had been reported by Earth-based observers but these features were so elusive that confirmation had to wait until the *Pioneer* and *Voyager* craft arrived in Saturn's vicinity. Four further rings have been found, and they have been labelled rings D, F, G and E. In increasing distance from Saturn the rings are: D, C, B, A, F, G and E (see Fig. 10.4).

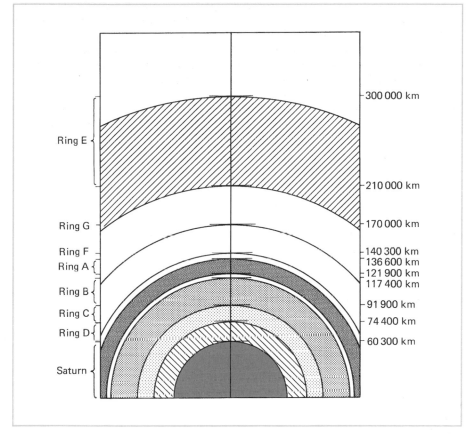

Figure 10.4 Saturn's ring system.

10.7 *Voyager* – a Detailed Look at the Rings

As *Voyager 1* approached Saturn in 1980, Cassini's Division and Encke's Division were visible from afar but, surprisingly, other divisions also became visible. When the probe was close to Saturn the rings were seen to be not bland sheets but rather composed of thousands of individual concentric ringlets. These give the rings a "gramophone record" appearance. We thought we understood Saturn's rings – how wrong we were!

Most of the individual ringlets are circular but some were found that are slightly eccentric. Also, the major divisions, such as Cassini's, are not empty of ring particles but have thinly populated ringlets of their own. One particularly peculiar feature is the F ring. Close-range photographs show this ring to be formed of three components that appear to be "braided" or "plaited", together with occasional bright knots along its length. Another puzzle!

Another surprise is the existence of radial "spokes", or dark, finger-like shadings, on the rings. Theoretically, any such features to form in the rings ought to be sheared by differential rotation, but the spokes preserve their identities for one or two rotations, though some shearing does occur in this time. One explanation for their existence is that very fine particles in the rings are electrically charged (by friction – in the same manner as terrestrial thunderstorm clouds) and are levitated above and below the ring plane. The precise mechanism that causes the levitation is uncertain, though Saturn's magnetic field is thought to be a factor.

The major divisions in Saturn's rings were formerly attributed to resonance effects with Saturn's extensive family of satellites. For instance, Cassini's Division is situated where a ring particle would have a period half that of one of Saturn's major moons, Mimas. A particle in this zone would experience a gravitational tug from Mimas twice every orbit. Gradually any particles in the zone would be swept into different orbits, leaving it particle-free. The *Voyager* results have not caused us to abandon this explanation, but other processes also occur to produce the thousands of ringlets observed.

Small *shepherding satellites* are thought to be responsible for the production of the ringlets. The general idea is illustrated in Fig. 10.5. In (a) the satellite S_1, which is moving more slowly than the particles in the ringlet, causes the outermost particles to be slowed down and to drop in orbital radius as a consequence. Meanwhile the

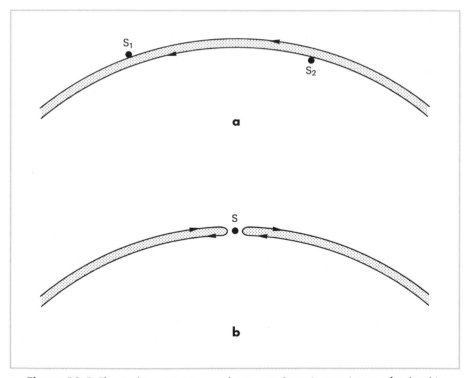

Figure 10.5 The ringlet-maintaining mechanisms in Saturn's rings (see text for details).

Figure 10.6 Saturn's ring system photographed by *Voyager 2*, showing the "grooved" appearance of the rings as well as the dark "spokes". JPL photograph.

satellite S_2 causes nearby ring particles to speed up and be driven into higher orbits. In this way the two satellites cause the particles to be confined in a narrow ringlet.

In (b) a satellite lies inside the ringlet and forces particles in higher and lower orbits to exchange momentum when they pass near the satellite. Eventually they settle into the ringlet. Both mechanisms may be in operation, in addition to the classical resonance theory, but this may not be the whole story. At present we do not know. Certainly a few of the shepherding satellites have been observed in the ring system. Figure 10.6 shows a close-range photograph of part of Saturn's ring system, obtained by *Voyager 2*.

10.8 The Formation of Planetary Rings

In 1848 the French mathematician Edouard Roche proved that any large solid or liquid body passing too close to a planet would be disrupted by shearing forces and spread around the planet as a ring of particles. The limiting distance from the planet has become known as the *Roche limit*, and its value depends upon the nature of the body. A strong body would survive intact until it approached to a closer distance than a weak body.

Alternatively, a ring of matter closer to a planet than the Roche limit could never form into a single body. So far we know of three planets in our Solar System which have rings – Jupiter, Saturn and Uranus. In each case the ring systems lie within the Roche limits of the parent bodies.

So, which process is responsible for the planetary rings – the destruction of a former body or the prevention of any body from forming? We cannot be absolutely certain, but in the case of Jupiter and Saturn it is thought likely that the rings originated from material left over from the formation of the planets themselves. This material, lying closer to the planet than the Roche limit, could never coalesce into a solid body. In the case of Uranus we are even less sure, because it appears that the rings are basically rocky in nature and are thus quite unlike the nature of Uranus itself. Perhaps this is a case where a former satellite, or some other passing body, entered too close and was torn asunder.

10.9 The Satellites of Saturn

Saturn has the most extensive satellite system of any planet in the Solar System. The first satellite to be discovered and, at magnitude 8.4, the brightest of Saturn's family, is Titan. Titan was found by Christian Huygens in 1665. In the late seventeenth century Cassini discovered Iapetus (magnitude 11.0), Rhea (magnitude 9.8), Dione (magnitude 10.4) and Tethys (magnitude 10.3).

William Herschel, using his great 48 inch reflector, found two more moons in 1789: Mimas (magnitude 12.1) and Enceladus (magnitude 11.8). William Bond added to the list with his discovery of Hyperion (magnitude 13.0) in 1848, while William Pickering found the faint (magnitude 16.5) Phoebe in 1898.

Aside from Titan, all the satellites are less easy to see than one might think, given their magnitudes. This is because they are immersed in the glare from Saturn itself. All the satellites are seen at their best when the rings are presented edge-on to us. It was at the edgewise opposition of 1966 that a new, faint satellite – Janus – was found to orbit close to Saturn. American astronomers found another satellite that shared Janus's orbit and in the edgewise opposition of 1980 a twelfth faint satellite was discovered. Thanks to the *Voyager* probes, we now know of more than twenty satellites in orbit about Saturn. Here, we discuss just the chief moons; Titan is considered separately.

Mimas

Mimas orbits Saturn at a mean distance of 185 000 km. It is approximately spherical with a diameter of only 390 km. We think that it is mainly composed of ices, and it has a mean density of 1200 kg/m³. The surface of the satellite, a pale brown covering of "dirty ice", is heavily cratered. It sports one massive crater that is more than a quarter of the diameter of the satellite itself. The impact that caused this crater must have come close to shattering the satellite entirely.

Enceladus

Enceladus orbits at a distance of 238 000 km from Saturn. It is, like Mimas, chiefly composed of ices. Its density is slightly less than Mimas: 1100 kg/m^3. Enceladus has a bright, whitish, suface that is partially cratered. There is evidence that ice flows have obliterated craters in large areas of its surface. Enceladus is rather larger than Mimas, having a diameter of 500 km.

Tethys

Tethys orbits beyond Enceladus, at a mean distance of 295 000 km from Saturn. It is even less dense than Enceladus, having a mean density of only 1000 kg/m^3. It, too, must be composed of ices. Tethys is rather larger than Enceladus, having a diameter of 1050 km. It has a pale brown and heavily cratered surface. One striking feature is an enormous rift valley that splits the surface, which perhaps was created when the subsurface fluids froze and expanded after the outer crust solidified soon after the formation of the satellite, over 4500 million years ago.

Dione

Dione orbits Saturn at a mean distance of 377 000 km. It is similar in size to Tethys, but obviously has a different composition, since its density is 1400 kg/m^3. Dione's rather greyish surface shows a distinctive pattern of bright radial streaks and areas of heavy cratering. It has an overall diameter of 1120 km.

Rhea

Rhea orbits Saturn at a mean distance of 527 000 km. It is larger than Dione, having a diameter of 1530 km, and is only slightly less dense. Rhea displays a highly reflective, pinkish surface, which is oddly different in each of its two hemispheres. One is heavily cratered but the other is rather bland, save for bright patches and streaks.

Hyperion

Hyperion orbits Saturn at a mean distance of 1 481 000 km. It is a small, oblate, body having a mean diameter of only 290 km. It possesses an icy and rather pitted surface.

Iapetus

Iapetus orbits Saturn at a mean distance of 3 560 000 km. It is much larger than Hyperion, having a diameter of 1440 km and a mean density of 1200 kg/m³. Iapetus is peculiar in having two distinctly different surface coverings. On its leading hemisphere as it orbits Saturn (it has a captured rotation) Iapetus is covered in a very dark, brownish material. The trailing hemisphere sports an icy surface nearly six times brighter.

Phoebe

Phoebe is the outermost known satellite of Saturn's retinue, orbiting at a mean distance of 12 930 000 km from the planet in a retrograde, highly inclined, and very eccentric orbit. Its diameter is only about 140 km, and it is thought to be a captured body, not an "original" satellite of Saturn.

10.10 Titan

Orbiting at a mean distance of 1 222 000 km from Saturn (between the orbits of Rhea and Hyperion), Titan is the biggest of Saturn's satellites. Its 5120 km diameter and 1900 kg/m³ density also makes it the most massive of the Saturnian moons.

Titan is the only satellite known to be surrounded by a substantial atmosphere. Its opaque, cloudy, mantle gives it a distinctly orange colour and a very bland visual appearance. The cloudy covering is so total that no surface details were visible to *Voyager 1*. This is a situation reminiscent of the planet Venus, though we must be careful to realise that the two worlds are totally different. For instance, the temperature on Titan is estimated to be a frigid –175 °C, compared with the fiery heat at Venus's surface.

A major surprise from the results of the *Voyager* programme was the discovery that the atmosphere of Titan is composed of about 90% nitrogen, with methane making up most of the rest. The atmospheric pressure at the surface of Titan is thought to be about 50% greater than that at the surface of the Earth.

Nitrogen and methane are both dissociated in Titan's upper atmosphere by solar radiation. As a result of chemical recombination, hydrogen cyanide and other more exotic organic compounds are formed. These compounds are thought to be reponsible for the orange colouration. Scientists have speculated that the conditions might be correct for a rain of liquid methane, laced with organic compounds, to fall on Titan's surface. Some think that Titan might have substantial oceans of liquid methane covering its frigid surface.

The surface is thought to be icy, but may be covered in brown, tarry deposits of the compounds washed down in the rain. In the year 2004 we may know for sure, as it is at this time the two-part space probe *Cassini-Huygens* should reach the Saturnian vicinity. If all goes well (it should be launched in 1997) it will drop a lander onto Titan's mysterious surface.

In biochemical terms, conditions on Titan may one day be suitable for the initiation of life, though not for thousands of millions of years. Then, the Sun may have increased its temperature sufficiently to allow some form of life to develop on this exotic world (by which time the Earth will be a roasted and dead globe in the fiery regions of the Solar System).

Questions

1 Describe the appearance of the planet Saturn as seen through a small telescope (say a 25 cm reflector), and very briefly explain the natures of the visible features.

2 Briefly describe the orbit, size and nature of the planet Saturn.

3 Explain why the rings of Saturn change their appearance as viewed from the Earth, from year to year.

4 Write an essay about the rings of Saturn (you need not explain why the rings change their angle of presentation to us on the Earth).

5 Write a short essay about the atmosphere of the planet Saturn.

6 Compare and contrast the Earth, the planet Jupiter and the planet Saturn.

7 Write an essay about the satellite family of the planet Saturn.

Chapter 11

The Twilight Zone of the Solar System

THE only planets known before the invention of the telescope were Mercury, Venus, Mars, Jupiter and Saturn. For centuries scholars thought that the Solar System consisted of just these five planets plus the Earth, the Sun and the Moon. Then, in 1781, the academic world received a rude shock when an amateur astronomer discovered another planet with his home-made reflecting telescope!

11.1 The Planet Uranus

Friedrich Wilhelm Herschel was born in Hanover in 1738. A talented member of a musical family, Herschel became a musician in the Hanoverian army. He found that a military life did not suit him and left the army and moved to England in 1757. He adopted the name of William Herschel and, later, his sister Caroline joined him in Georgian England. William obtained various musical posts and eventually settled in the little spa town of Bath, where he became an organist at the local chapel. Ever since his youth Herschel had been interested in science, and his more settled life in Bath allowed him to devote his spare time to scientific study and experiment.

In particular, Herschel's fascination with astronomy prospered, but the prices of contemporary telescopes were too high for his salary and so he set about making his own. After many failures, he eventually made a usable one. His energy and determination were outstanding and he ultimately became the maker of the highest-quality telescopes of his time. Herschel did not stop at constructing telescopes; he used them for studying the heavens and in 1781 made an accidental discovery that was to change his whole life, though he did not realise its true nature at the time he made it. He wrote a paper, "Account of a Comet", which was communicated to the Royal Society by his friend, Dr William Watson. The first part of the account read:

> On Tuesday the 13th of March, between ten and eleven in the evening, while I was examining the small stars in the neighbourhood of H Geminorum, I perceived one

that was visibly larger than the rest: being struck by its uncommon magnitude, I compared it to H Geminorum and the small star in the quartile between Auriga and Gemini, and finding it to be so much larger than either of them, suspected it to be a comet.

Herschel was using a 6.2 inch (157 mm) Newtonian reflector of his own construction. Its speculum metal mirror had a focal length of 85.2 inches (2.16 m). The magnification which Herschel was using when he first saw Uranus was ×227. Uranus is of the sixth magnitude and subtends an apparent diameter of only four arcseconds. The fact the Herschel noticed at a glance the non-stellar nature of this little disk is testimony to the superiority of his telescope over the other instruments of his day.

It soon became apparent to astronomers that Herschel's object did not follow the path of a comet; its motions revealed it to be a planet orbiting the Sun at twice the distance of Saturn. Herschel immediately became famous. He was knighted and appointed astronomer to King George III. The small bursary that went with this position allowed him to devote much more time to his astronomical interests. In gratitude Herschel named his planet "Georgium Sidus" (George's Star) but the astronomer Lalande suggested the name "Herschel" and Johannes Bode suggested "Uranus". The latter, mythological, name was felt to be in keeping with those of the other planets and came into common use.

Uranus moves in an orbit 2900 million km from the Sun, with a sidereal period of 84 years. Owing to the long orbital period of the planet, it comes into opposition only about $4\frac{1}{2}$ days later each year. Uranus's diameter is 52 300 km and its mass is $14\frac{1}{2}$ times that of the Earth. This gives it a mean density of about 1200 kg/m^3.

Seen in a telescope, Uranus's small, bluish-green disk is devoid of any detail. Its upper atmosphere is very clear, mainly composed of hydrogen, helium and methane. It is the selective absorption of sunlight by methane that gives rise to the colour of the planet. Infrared measures indicate that the temperature of the lower atmosphere is a frigid –220 °C. The planet is thought to have a partially molten rock core overlaid by a deep "ocean" of water, perhaps containing some liquid ammonia and methane, over which is the atmosphere. The conditions deep within the liquid mantle must force the water into its "metallic", electrically conductive, state.

Until the space probe visit by *Voyager 2*, five moons were known to orbit Uranus. In increasing distance from the planet they are: Miranda, Ariel, Umbriel, Titania and Oberon. The brightest of these, Titania and Oberon, were discovered by Herschel in 1787. They are all small, icy bodies.

One major oddity of Uranus is that its rotation axis is much more inclined than any of the other planets. In fact, it is inclined by 98° to the perpendicular to the plane of its orbit. So, Uranus technically spins in the retrograde direction! A consequence is that, despite the $15\frac{1}{2}$ hour rotation rate of the planet, the poles of the planet experience a day and night each 42 years long. From the Earth we see first the south pole presented to us, then the equator, then the north pole, then the equator and then the south pole again, over a cycle of 84 years.

Astronomers are interested when the motions of a planet cause it to occult a star, since the way the light changes upon immersion and egress gives accurate information on the atmosphere of a planet as well as the planet's diameter. With these objectives in mind, astronomers observed the occultation of the star SAO 158687 in March 1977. Observation groups were set up at various locations all over the world to observe the event. Most of these groups were successful. Indeed, the astronomers received an added bonus. The star was seen to flicker and dim a number of times before the disk of Uranus occulted it. The flickering was repeated some while after the main disk of Uranus cleared the star. A set of rings had been discovered to encircle the planet in its equatorial plane.

Nine ring components were identified, all thin and narrow; the innermost ring has a radius of 84 000 km, while the outermost has a radius of 102 000 km. Subsequent infrared images obtained from the Earth, though poorly resolved, do show the rings. The fact that they do not show up in visible light indicates that the rings are made of a very dark material. We were to learn much more about Uranus and its rings when *Voyager 2* reached the planet in January 1986.

11.2 *Voyager* to Uranus

As the probe approached Uranus it became apparent that the cyan disk of the planet was extremely uniform and featureless in appearance. Subjecting the *Voyager* images to a large amount of computer enhancement revealed some detail. It was found that the sunlit pole (at the time of encounter the planet's south pole was pointed sunward) is very slightly redder and more reflective than the equatorial regions. This might be due to a haze of hydrocarbon molecules, such as acetylene (ethyne, C_2H_2). Seen at middle latitudes, rotating round the planet with periods of 15 to $16\frac{1}{4}$ hours, were a few vague clouds which scientists think might have been formed from simple organic compounds like acetylene.

Project scientists discovered a further, tenth, ring in the *Voyager* images. This ring is situated between the outermost pair previously known. Also, ten minor satellites were discovered, to add to the five moons already known. The Uranian rings appear to be made of large chunks of dark rock, averaging about 1 m in diameter. Some fine ring material was found, but much less than scientists had expected. Over a period of time the ring particles should grind against each other and release much fine material into the ring plane. Another puzzle lies in the rings themselves. They are all extremely narrow and thin. What keeps them like this against their natural tendencies to spread? Only two of the newly discovered satellites appear to act as shepherding moons. A most unusual phenomenon.

The *Voyager* probe detectors found that Uranus's magnetic field is dipolar in form, like that of the other planets. The magnetic polarity is in the same sense as Jupiter's and Saturn's fields – in other words, magnetic north and south are inverted by comparison with the Earth's field. The dipole moment of Uranus's magnetic field is roughly a tenth that of Saturn, though the flux density at the surface of Uranus is a little stronger than that at the surface of Saturn. The magnetosphere extends for about nine Uranus diameters in the Sunward direction.

One of the biggest surprises of all is that the magnetic axis is inclined at 60° to the rotation axis. In no other planet is the magnetic tilt so large. Remember, also, that the rotation axis is canted over at more than a right angle. Neither does the magnetic axis pass through the centre of the planet; it is, instead, offset by thousands of kilometres. The situation is illustrated in Fig. 11.1.

The rotation period of the magnetic field, thought to be the same as that of the planet's interior, was measured at $17\frac{1}{4}$ hours. This is substantially longer than the rotation of its atmosphere, and means that the winds on the planet always blow from the direction of rotation. Previous knowledge of atmosphere dynamics had suggested the opposite would be the case, given that the poles receive more sunlight than the equatorial regions. Certainly the values of temperature and temperature gradients within the atmosphere are not what had been predicted. All this means that unknown processes must be going on in the Uranian environment.

Another peculiarity is that Uranus is immersed in a vast cloud of hydrogen gas. This gas appears to be responsible for an aurora-type glow in ultraviolet wavelengths, the *electroglow*, covering just the sunlit hemisphere of the planet. It takes place between about 1000 and 2000 km above the visible atmospheric layers. The glow appears to be caused by high-energy electrons colliding with the hydrogen atoms. Perhaps these electrons result from some of the hydrogen atoms being split into their component electrons and protons by the short-wave radiations from the Sun. This would explain their source, but not how they are accelerated to high kinetic energies. This is yet another puzzle to add to the list!

Recent (1994–5) Hubble Space Telescope images of Uranus have shown white clouds formed at mid-southern latitudes. Clearly Uranus's bland appearance is not the permanent situation. It may well be significant that at the time of the HST images the Sun was overhead from the mid-southern latitudes, rather than over the south pole, as it was at the time of the *Voyager* encounter. In effect, Uranus's meteorological activity may very much depend on its seasons.

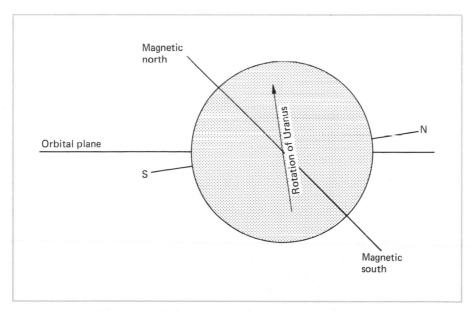

Figure 11.1 The rotation and magnetic axes of Uranus.

11.3 The Satellites of Uranus

Voyager 2 also obtained close-up views of Uranus's five major moons, as well as discovering a further ten minor moonlets. None of the satellites are larger than 1600 km across and, in the intense cold nearly three billion kilometres from the Sun, little was expected in the way of evidence of geological activity. Yet again the planetary scientists were to be surprised.

Oberon

Oberon's mottled and cratered surface displayed many bright craters surrounded by bright ray systems. Many of the crater floors had dark patches in them, as if "dirty water" had erupted in an icy form of volcanism. This satellite is the outermost of Uranus's retinue and is the second largest, with a diameter of 1560 km. Its orbital radius is 586 000 km.

Titania

Titania is the largest moon, with a diameter of 1590 km. This satellite orbits at 438 000 km from Uranus. The Voyager cameras showed Titania to have a pockmarked surface crossed by great canyons and valleys.

Umbriel

Umbriel, orbiting 267 000 km from Uranus, has a strangely dark covering over its entire 1190 km diameter globe. Ancient-looking craters abound on its surface. At least this satellite seems to have the expected quiescent surface.

Ariel

Ariel produced universal surprise when the first of the *Voyager* pictures arrived after their $2\frac{3}{4}$ hour journey to the Earth (radiowaves take this long to cross the immense distance between Uranus and the Earth). Its 1170 km diameter globe was seen to be covered in a network of colossal grooves and valleys. This satellite moves round Uranus at a distance of 192 000 km.

Miranda

Miranda, 130 000 km from Uranus, is the most astounding of the Uranian satellites. Its bizarre surface looks like a patchwork quilt of different sorts of terrain (see Fig. 11.2). Patterns of deep grooves swirl across its surface in several places.

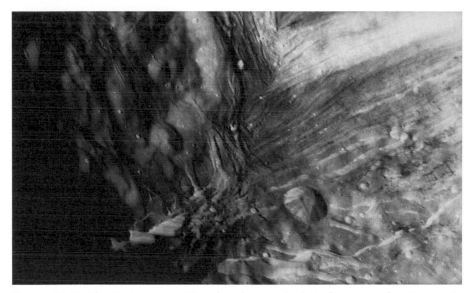

Figure 11.2 Chaotic terrain on Miranda, photographed at close range by *Voyager 2*. Photograph courtesy JPL.

In others there are no grooves but just rough and hummocky ground covered in craters. Enormous, jagged cliffs soar up from its surface in the borders between the different types of terrain. Miranda is only a small body (about 480 km across), but its surface speaks of past activity on a fantastic scale.

Some scientists think that Miranda has been smashed into several large pieces by a collision, perhaps with another moon or an asteroid, and has later reassembled into its present form. Perhaps some past cataclysmic event is responsible for the appearances of Miranda and Ariel (and perhaps also the other moons), as well as the strange axial tilt of Uranus and the inclination of its magnetic field. *Voyager 2* has told us much about Uranus, including that we still have much to learn.

11.4 The Planet Neptune

Positional observations of the planet Uranus soon after its discovery revealed unaccountable perturbations. Uranus seemed to drift off its predicted position. Some astronomers began to wonder if another large body, as yet unseen, was pulling Uranus off course.

A brilliant young mathematician, John Couch Adams, began work on the problem early in the 1840s. In 1845 he enlisted James Challis, an astronomer at Cambridge University to help him approach the Astronomer Royal, George Airy. Airy was known as a petulant character who had little time for the young, and Adams's attempts to secure an audience with him proved fruitless. Nevertheless, Adams did send his results to Airy but, characteristically, Airy did nothing positive with them.

Meanwhile, in France, Urbain Jean Joseph Le Verrier, a chemist turned astronomer, came up with a similar result to Adams. Le Verrier's work was published. At long last Airy enlisted Challis to do a search for the planet from Cambridge, but Challis's mode of working was slow and inefficient, even if it was meticulous. Le Verrier wrote to the astronomers at the Paris Observatory but, getting no positive response from them, he wrote to Johann Galle at the Berlin Observatory. On 23 September 1846 Galle and a student, Heinrich d'Arrest, found Neptune with the Observatory's 9 inch (230 mm) refractor. This was the first night of their search and Neptune was found less than a degree away from its predicted position!

Uranus had been a chance discovery, but Neptune was tracked down using the power of mathematics. It was a notable achievement and both Adams and Le Verrier received well deserved credit – and Airy received well-deserved criticism! Neptune appears to be similar to Uranus. Neptune is of the seventh magnitude and so a telescope or a pair of binoculars is needed to show it at all times. It shows a pale blue disk through a telescope, about 2 arc seconds across. Neptune moves at a mean distance of 4500 million km from the Sun, taking 165 years to complete one orbit.

In many ways Neptune and Uranus are rather similar. Neptune's diameter is 49 500 km, which is a little less than Uranus's, but its mass is larger, at 17 times that of the Earth. Thus the average density of Neptune is rather higher than that of Uranus – 1660 kg/m^3. Infrared measurements have indicated that the atmosphere of Neptune has a similar temperature to that of Uranus. Again like Uranus, its blue colour is chiefly due to the absorption of red light by methane gas. In common with Jupiter and Saturn (but unlike Uranus), Neptune radiates away into space more heat than it receives from the Sun. In fact it radiates more than $2\frac{1}{2}$ times the received solar heat energy, more than that of either Jupiter or Saturn.

It is not thought that the heat output is supplied by an internal rain of helium, as is the main agency at work in Jupiter and Saturn. Instead the slow release of its primordial heat of formation is thought to be the sole source (remember Neptune is much further out from the Sun, and so we considering a much smaller heat exchange). However, the absence of a net outflow of heat from Uranus is a puzzle. It probably gives us clues to the differences in internal structures of the planets.

No atmospheric features had ever been unambiguously observed from earth in visible wavelengths, owing to the difficulty of observing Neptune's tiny disk through our unsteady atmosphere. However, infrared studies of the planet have shown that its atmosphere is not clear but clouded. Although poorly resolved, the clouds were seen just about well enough to enable determinations of the rotation period of the planet to be made (at least for a limited range of latitudes). At a latitude of 30° S the rotation period was found to be 17.7 hours and 17 hours at 38° S. Our vision of the planet was to sharpen considerably once Voyager 2 reached it in 1989.

11.5 *Voyager* at Neptune

Far from being like the bland Uranus, Neptune surprised everybody by displaying a much more dynamic visage as Voyager 2 swept in for its encounter in August 1989 (see Fig. 11.3). The globe was seen to be streaked with white clouds of

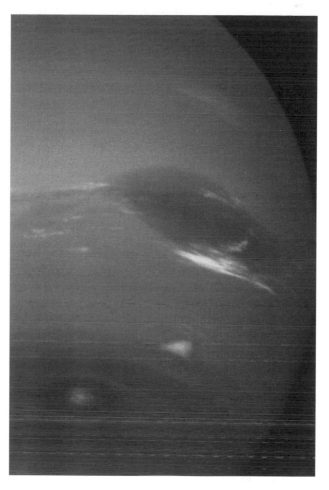

Figure 11.3 Neptune from *Voyager 2*. The Great Dark Spot is especially prominent as are white streaky clouds of methane crystals and a smaller spot. Photograph courtesy JPL.

methane crystals, which float in the clear atmosphere well above a layer of chemical haze (possibly a mixture of hydrogen sulphide and ammonia), which defines the visible "surface".

Particularly striking were the dark spots, especially the Great Dark Spot well shown in Fig. 11.3. These are now thought to be clearings in Neptune's atmosphere which allow us to see to deeper levels. The circulating motions of clouds round the spots reveal vast wind-shear forces, like those around Jupiter's GRS. Recent HST images are well enough resolved to show that the Great Dark Spot has now disappeared. This certainly was not expected.

Winds speeds proved to be very high compared with the basic rotation period of the main bulk of the planet, as is the case for the other gas-giants. Winds up to 300 m/s are typical, blowing in a direction opposite to the rotation of the planet. As can be seen from the clouds and spots, regions of shear between wind streams abound. Owing to Neptune's outpouring of heat, the temperatures in its lower atmosphere are broadly similar to those of Uranus, despite Neptune's greater distance from the Sun.

The preliminary earth-based radio observations plus the results from the *Voyager* probe have enabled scientists to build up a picture of the planet's magnetosphere. Neptune, like Uranus, has a broadly dipolar magnetic field (though rather more chaotic near the surface of the planet), the axis of which is both heavily inclined (at 47 degrees to its rotation axis) and offset from the centre of the planet. In the case of Neptune, the offset is by a full 55% of the radius of the planet. So, the source of Neptune's magnetism is certainly not its core. Instead, the magnetic field must originate from motions in the aqueous mantle that is thought to surround the possibly solid core of the planet. The magnetic field is a little larger than that of Uranus and is orientated in the same sense as for the other gas-giant planets – the opposite of that of the Earth.

From the radio waves generated by the magnetic field as it rotates, it has been determined that the field (and hence the inner region of the planet) turn with a period of 16 hours 7 minutes. Neptune's rotation axis is inclined at about 29° to the perpendicular to its orbital plane.

11.6 Neptune's Rings and Satellites

The *Voyager* encounter revealed that Neptune has a very faint, six-component ring system (first indications of them were observed from Earth in 1980). The rings span a radius of about 40 000 km to about 70 000 km, and are thought to be composed of fragments of one or more disintegrated satellites. The rings are probably transient in nature and the present system may be less than 500 million years old.

Voyager discovered six tiny (less than 420 km across) moonlets orbiting within the ring system. Beyond them are the two major moons, Triton and Neried, both long-known by astronomers on the Earth.

Triton moves around Neptune, at a distance of 353 000 km, with a nearly circular, but retrograde, orbit. The orbit appears to be decaying, so that Triton will cross the Roche limit in about 100 million years. If the ring system is faint at the moment, it certainly will not be after Triton has been broken up and the fragments disseminated around Neptune! *Voyager 2* showed that Triton is a strange globe (diameter about 2600 km) of pinkish and bluish areas. It has an extremely rarified nitrogen atmosphere (with some methane), and great nitrogen-driven geysers of dark material erupt onto it its patchwork surface of bluish- and pinkish-tinted ices. Triton is thus the third orb in the Solar System to sport active volcanism.

The other satellite, Nereid, moves in a direct but highly eccentric orbit, with a mean distance of 5.6 million km from Neptune. Nereid is an irregular lump of dirty ice, about 170 km across its longest diameter and has an albedo of about 0.12.

11.7 The Planet Pluto

After the discovery of Neptune, speculation was rife as to the possibility of another planet even further out in the Solar System. Could it be found? It was bound to be faint, and would certainly move very slowly against the backdrop of the stars. Theorists searched for further discrepancies in the motions of Uranus (Neptune had not been observed for a long enough period to have moved very much). Prominent in the quest was Percival Lowell, who was already famous in connection with the Martian "canals".

Lowell began trying to calculate the position of the proposed planet, and started a photographic hunt in 1905. His method was to take several photographs of the region of sky in which he expected to find the object. These photographs were spread over several days, or weeks, and they were all compared using a device known as a *blink comparator*. In this device two plates of the same region of sky are mounted and viewed alternately using an optical system. Any differences between the plates then become obvious. A planet would betray itself because of its movement. Unfortunately, Lowell died in 1916 and the search was suspended.

The planet search at the Lowell Observatory was resumed in 1929 when Clyde Tombaugh was hired to use a newly constructed photographic telescope specifically for finding the object. By setting out on an exhaustive programme of photographing regions of the sky and using the blink comparator, he eventually tracked the planet down and its discovery was announced in March 1930. Tombaugh and the Lowell Observatory became world-famous, and the Solar System was then known to have nine major planets.

It is fair to say that the discovery of Pluto raised a number of problems. It was much fainter than expected. It appears as a star-like point of the fourteenth magnitude even in the largest telescopes. Only by assigning an impossibly great density to Pluto could its mass be sufficient to explain the perturbations in the orbit of Uranus which led to the search. Pluto's orbit is highly irregular. The planet orbits at a mean distance of 5900 million km, from the Sun, but this distance can vary from 4400 million km to 7400 million km. At certain times in Pluto's 248 year orbit it comes nearer to the Sun than does the planet Neptune! Further, Pluto's orbit is greatly inclined to the ecliptic plane (17°.2). Some astronomers have speculated that Pluto might be an escaped satellite of Neptune, though this idea is now rather out of fashion. Whatever the origin of Pluto, many now think that it does not warrant the title of "major planet".

Pluto's diameter is roughly 2300 km and spectra indicate that its surface is covered in frozen methane. From Pluto the Sun appears no more than a brilliant star and the temperature on the surface of the planet is certainly no higher than $-230\,°C$.

In 1978 James Christy, using the 1.54 m reflector of the US Naval Observatory at its Flagstaff, Arizona, station, photographed the planet in order to try to measure its diameter. The very best photographs show a "bump" attached to the image of Pluto – an attendant satellite had been found The satellite is about a third of the size of Pluto itself and orbits at about 17 000 km from it with a period of 6.3 days. Pluto also rotates on its axis with this period. Thus the satellite (named Charon) appears fixed in the sky when viewed from one hemisphere of Pluto and is always invisible from the other hemisphere. The motions of the satellite allow a calculation of Pluto's mass. It proves to be $\frac{1}{380}$ of the Earth's mass.

The Hubble Space Telescope has turned its sharp vision towards Pluto and images have been obtained which show light and dark markings on the globe of the planet, as well as showing Charon very clearly separated from Pluto. Perhaps Pluto has a similar surface to Triton. Certainly, astronomers think that the surface is largely covered in methane ice, "dirtied" with organic compounds, and the density of Triton and Pluto are similar at about 2000 to 2100 kg/m^3.

A very rarified atmosphere of methane, with traces of nitrogen, carbon monoxide and neon, has been detected, but this may well freeze out onto the surface at the times when Pluto is at aphelion.

So the question remains: is there another major planet orbiting at a still greater distance from the Sun? At the moment it must be admitted that we do not know. Indeed, centuries may pass before we do.

Questions

1 Write an account of the discoveries of the planets Uranus, Neptune and Pluto.

2 Write an essay about the planet Uranus and its satellites, including the main facts learned from *Voyager 2*.

3 Give the main facts known about the planet Neptune and its satellite family, including the main facts learned from *Voyager 2*.

4 "Pluto is the outermost planet in the Solar System." Consider this statement and discuss whether it is necessarily true.

5 Write an essay about the planet Pluto, describing how it was discovered and what facts have been learned about this planet. Remember to include notes about any satellites this planet may have.

6 Briefly discuss the evidence for a planet orbiting the Sun at a greater distance than Neptune or Pluto.

Chapter 12

The Debris of the Solar System

S O far we have been considering the major bodies of the Solar System. There is also a managerie of smaller objects. Only recently have astronomers been able to begin the task of unravelling the history of the Solar System and to fit the great variety of objects into the story of its early formation. Our ideas are still rather sketchy, but in their basic outline are unlikely to be very far wrong.

12.1 The Formation of the Solar System

The Solar System initially condensed from a massive cloud of interstellar gas and dust about 4700 million years ago. The mass of the initial cloud was thought to be about a thousand times bigger than that of our Sun. It was composed mainly of hydrogen gas and had a temperature of about –260 °C. Traces of heavier elements enriched the gas. These had been synthesised from hydrogen and spattered into space by the violent death throes of earlier stars.

Instabilities caused the major cloud to split into separate cloudlets. As each of these cloudlets began to contract under its own gravitation so the gas became hotter. (It is a well-demonstrated effect that when a gas is compressed its temperature increases.) Also, the cloudlets began to develop vortex motions (again a well-demonstrated phenomenon). As the cloudlets further contracted so the conservation of angular momentum caused them to rotate faster and become flattened out into disks (see Figure 12.1, *overleaf*). We will now concentrate on the condensing cloudlet that gave birth to our Solar System.

Much of the mass of our disk-shaped protosolar cloud was concentrated into its central bulge. The temperature varied from a little above absolute zero (which is –273 °C) near the edge of the disk to several thousands of degrees near the centre of the bulge. As the central regions of the cloud continued contracting, so the temperature further increased to millions of degrees. Under these conditions nuclear reactions began at the centre of the central condensation – the Sun began to shine.

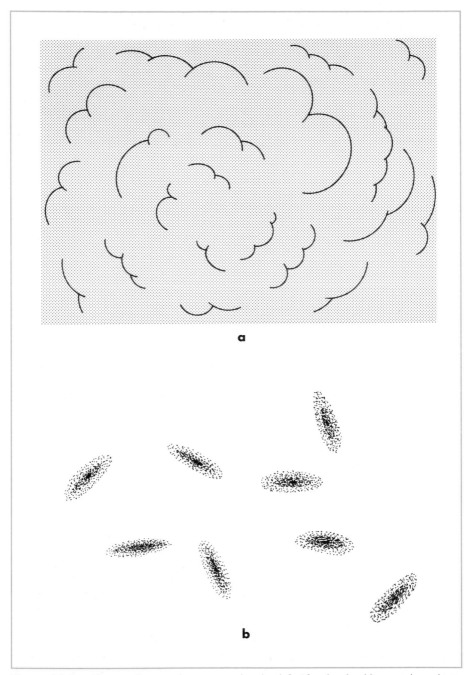

Figure 12.1 a The initially amorphous protosolar cloud. **b** After the cloud has condensed into disk-shaped cloudlets.

Meanwhile the disk of material surrounding the young Sun itself began to fragment and condense into rings of material. The major rings each gave birth to a planet, with much assorted debris remaining. The shining Sun effectively blew away a lot of this residue of gas and fine dust, leaving the major bodies and the larger debris.

This scenario explains why all the major planets orbit the Sun in virtually the same plane and in the same direction. One would also expect the main concentrations of more volatile elements and compounds to lie in the outer regions of the Solar System, with the more refractory materials in evidence close to the Sun. This is exactly what is found in the Solar System today. The reason is that the temperatures in the inner parts of the *solar nebula*, the initial cloud of matter that formed the Solar System, were too great to allow appreciable amounts of the most volatile materials to condense in these regions. The inner, terrestrial, planets are mainly composed of high-melting-point materials, such as iron and silicates. The gas-giant worlds are chiefly made up of hydrogen and helium. The continuous shining of the Sun has also augmented the chemical differentiation within the Solar System by driving off the residual amounts of volatiles from the planets near the Sun.

Meanwhile, the other cloudlets also gave birth to stars, many of which must have planetary systems. All having been formed in the same region of space, the stars must have formed a vast cluster. Over the aeons the individual stars gradually dispersed, as a consequence of their own motions. There is little evidence of this clustering today, though other clusters of young stars can be seen. One example is the Pleiades, or Seven Sisters, in the constellation of Taurus. In this cluster we even see nebulous material, left over from the formation of the group, lacing the space between the stars. (That said, one piece of research work now indicates that the nebulosity in the Pleiades may *not* be the original, but merely some more material that is sweeping through the system.)

12.2 Planetary Cratering

Since the dawn of the telescopic era numerous theories have been put forward to explain the cratered surface of the Moon, and we have now observed that all the well-preserved surfaces in the Solar System (those of the quiescent rocky and icy planets and satellites) are extensively cratered.

Today it is thought that the craters were formed by the planets sweeping up much of the debris left over from the Solar System's formation. The planets would have been intensely bombarded by particles of rocky and icy material, ranging in size from the microscopic to vast bodies hundreds of kilometres across. The main cratering process was over by about 3500 million years ago. However, a declining rate of activity continued and continues even today.

In the case of our Earth, weathering and geological processes have all but removed the record of the early bombardment, though a few of the more recent craters still exist. One example is the famous Meteor Crater in Arizona, USA. This is a basin 170 m deep and of 1.2 km overall diameter. More is said about interplanetary material striking our planet later on in this chapter.

Table 12.1 Bode's law

Planet	Bode's law distance	Actual distance
Mercury	4	3.9
Venus	7	7.2
Earth	10	10.0
Mars	16	15.2
—	28	—
Jupiter	52	52.0
Saturn	100	95.4

12.3 Bode's Law

In 1766, Johann Titius published a numerical observation of the relative distances of the planets from the Sun. Basically, if we take the sequence of numbers 0, 3, 6, 12, 24, 48 and 96 and add 4 to each number the sequence becomes 4, 7, 10, 16, 28, 52 and 100. Titius set the distance of Saturn from the Sun as 100 units (Saturn was the outermost planet known in 1766) and found that the other planets very closely followed the sequence in terms of their relative distances (see Table 12.1).

Titius considered that the sequence fitted too well to be mere coincidence and he even proposed that the gap at relative distance 28 contained an undiscovered satellite.

Little notice was taken of Titius's theory until 1772, when Johann Bode republished it (without giving due credit to Titius) in a modified form. Bode proposed that an undiscovered major body occupied the orbit at 28 units from the Sun (clearly the satellite hypothesis of Titius was wrong since, by definition, satellites need a major body about which to orbit).

The relation became well known after Bode published it and we now know it as *Bode's law*. Table 12.1 gives the "law" in the form published by Bode, with the Earth's distance (10 units) normalising the sequence. The mean Earth–Sun distance is known as 1 astronomical unit (1 AU), and so all one has to do is to divide the Bode distance by 10 to express it in astronomical units – thus Jupiter orbits the Sun at a mean distance of 5.2 AU.

When Uranus was discovered in 1781, it was found to have a mean orbital radius of 192 Bode units. By Bode's law it should orbit the Sun at a mean distance of 196 units. The discrepancy is so small that astronomers became convinced that Bode's law had real significance. However, what of the gap at 28 Bode units (2.8 AU)?

12.4 The Celestial Police

In 1800 a group of six German astronomers decided that they would systematically search for a planetary body to fill the Bode gap. Among them were Johann Schröter, Franz Xaver von Zach, Karl Harding and Heinrich Olbers. The group divided the zodiac into a number of zones and their secretary, von Zach, contacted other astronomers in an effort to persuade them to join in the search. The idea was that each zone should have at least one astronomer patrolling it. The six German astronomers were nicknamed the "Celestial Police".

In fact the discovery was made by the Italian Guiseppi Piazzi, at the Palermo Observatory in Sicily, just before he received a request to observe from von Zach. On the evening of 1 January 1801 Piazzi, while engaged in the observational work necessary for the compilation of a new star catalogue, found an unfamiliar object in the constellation of Taurus. The object appeared to be like a star of the eighth magnitude. When Piazzi observed on the following night he found that it had shifted against the pattern of fixed stars, thus proving it to be a member of the Solar System.

12.5 The Asteroids

The mathematician Karl Gauss found that the new body orbited the Sun at a distance of 27.7 Bode units – close to what was expected. Clearly the new body, which Piazzi named Ceres, is very small. We now know that its diameter is about 1000 km. The Celestial Police continued the search, hoping to find further bodies in this region of the Solar System. On 28 March 1802 Olbers found a second small body, Pallas, which moves at about the same distance from the Sun but in a much more eccentric and highly inclined orbit than Ceres. In September 1804 Harding discovered a third, Juno, and in March 1807 Olbers found a fourth, Vesta. All these bodies have proved to be smaller than Ceres.

The four bodies became known as *minor planets,* or *asteroids.* The "Celestial Police" ceased their work in 1815 but in the following years other asteroids were discovered. Indeed, the application of photography caused the discovery rate to increase rapidly. Today more than two thousand of them are known.

Ceres remains the largest asteroid discovered. Most of them have orbital radii of between 2.5 and 3.5 AU. Many have very elliptical and highly inclined orbits. Indeed, several have passed very close to the Earth on various occasions. Such was the case with Hermes which, in 1937, came within 800 000 km of Earth, producing the inevitable "Earth narrowly avoids disaster" reports in the Press of the time.

The minor planet Icarus passes within 40 million km of the Sun. When at perihelion it becomes red-hot. Another, Hildago, reaches out nearly to the orbit of Saturn. Two small groups of asteroids, the Trojans, are trapped in the orbit of Jupiter. They occupy two regions of gravitational stability, one 60° in front of and the other 60° behind Jupiter itself. Jupiter is also responsible for several zones of avoidance in the Solar System, known as *Kirkwood gaps.* These are at radii from the Sun where any object would have an orbital period that is a simple ratio of Jupiter's.

Charles Kowal has found one asteroidal body that orbits between Saturn and Uranus. This object, which has been named Chiron (not to be confused with Charon, Pluto's satellite), could well be just one member of another asteroid belt in the outer Solar System. However, there is growing evidence that it is an icy body with a diameter of a few hundred kilometres. Traces of a gaseous mantle have been detected and this would put it in the class of the comets (comets are described in the next chapter).

The asteroids seem to be small rocky bodies. We only know of seven that are definitely larger than 300 km in diameter and the smaller examples seem to be highly irregular in shape (we can tell this by the way their apparent brightness changes as they rotate). A few of them are known to be double, or multiple, systems orbiting around their barycentres.

Scientists currently think that the asteroids originated from material that could never coalesce to form a major planet. The massive planet Jupiter would have caused tidal disruption of the ring of material that could have given rise to such a planet. A few small (Ceres-sized) planetoids probably did form, but collisions between these bodies caused them to fragment into innumerable pieces. To some extent this fragmentation process continues today, supplying the Solar System with fresh dusty and grainy material. Some of this matter intersects the orbits of the planets and eventually impacts upon their surfaces or burns up in their atmospheres in a brief flash of glory – a shooting star.

12.6 Meteors

Most people have seen shooting stars or at least have heard of them, yet relatively few know what they really are. A shooting star, or *meteor*, is a small piece of cosmic debris that enters the atmosphere at high speed, burning up by frictional heating and producing a bright trail as it goes. The particles may be rocky or icy in composition and they usually move around the Sun in elliptical orbits, either singly or in a swarm. If such a particle crosses the Earth's orbit just as the Earth passes that point then it will enter the atmosphere, usually at a very high relative speed. This speed can be as high as 80 km/s. Below a height of about 200 km the atmosphere is dense enough to reduce the speed of the particle. In this way kinetic energy is converted to heat energy and the temperature of the particle rises to thousands of degrees as a consequence. The high temperature causes ionisation of the air, and this produces the brilliant light which is visible from the ground.

A particle the size of a grain of sand will produce a very bright shooting star, while a pebble-sized meteor will result in a brilliant fireball accompanied by a bright flaring trail that persists for several seconds. Some of the brighter meteors are distinctly coloured and many pulse, or "sputter", in brightness before they are extinguished. The brilliant fireballs are rather rare. I have seen only a few notable examples over the years but a great many, probably hundreds, of the fainter ones, though I am not a meteor observer.

Micrometeorites continually rain down on the Earth. They are tiny granules, the largest of which are smaller than grains of sand. They are easiest to find deep on the ocean bed or in places like Antarctica, simply because they can lie undisturbed for thousands, or even (frozen in ice) millions of years. Their extraterrestrial origin can be ascertained by laboratory determinations of the ratios of certain isotopes they contain – a sort of chemical fingerprint which can distinguish them from earthly rocks.

Some meteors appear singly and from any direction in the sky. These are termed *sporadic meteors*. They also occur in shoals and these are termed *shower meteors*. In the case of shower meteors, the particles producing the shower are spread around an elliptical orbit and the Earth then intersects the stream of particles once a year to produce an annual meteor shower. There are many such showers that occur with great regularity each year. Since the particles of a given shower will be entering the Earth's atmosphere in approximately the same direction, they appear to emanate from a very definite small patch of sky, termed the *radiant*. This is purely a perspective effect, since the particles of the shower all enter approximately parallel to each other. Figure 12.2 illustrates the principle.

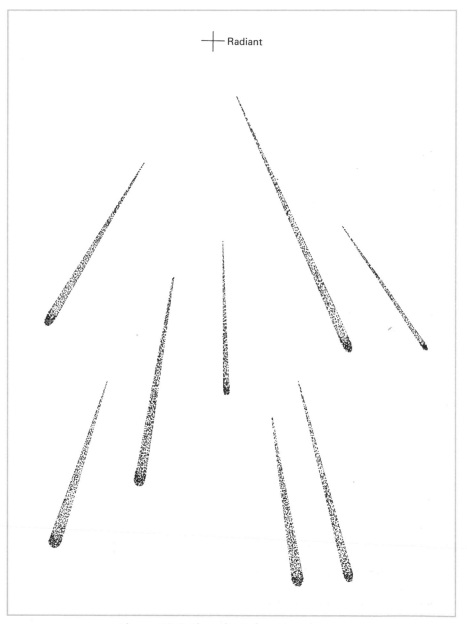

Figure 12.2 The radiant of a meteor shower.

For instance, the shower that occurs annually between 27 July and 17 August has its radiant in the constellation of Perseus. That is not to say that the meteors all appear in this constellation but rather that if all of the directions of the meteors are plotted on a star map and produced backwards, then all the tracks appear to intersect closely in Perseus.

For this reason, the shower is known by the name of the Perseid. The other annual showers are each denoted by the name of the constellation in which the radiant lies. Thus the shower that occurs between 15 October and 25 October is known as the Orionid – and so forth. Not all the annual showers are as intense as one another. The characteristics of the meteors of a particular shower vary quite considerably from one shower to another.

Also, successive showers of a particular annual display may, in some cases, be different in richness (the number of meteors falling in a given time) from one to another. For instance, the Perseid shower tends to produce a steady stream of meteors from year to year. This indicates that the material that gives rise to this shower is evenly spread around its orbit. However, in the case of the Leonid shower, the activity varies quite considerably from one year to the next. In this case the material is "bunched up" in certain places around the orbit. This point is illustrated, though not to scale, in Fig. 12.3.

As a consequence of the rotation of the Earth, meteors are roughly twice as common and faster moving in the hours after midnight than in the hours before. Before midnight a particular location on the Earth's surface is on the "trailing face" of the Earth and the meteors have to "catch Earth up", so to speak. After midnight Earth meets the meteors "head-on".

12.7 Meteorites

Occasionally a projectile enters our atmosphere that is large enough to survive at least partially intact and reach the ground. Such a body is then termed a *meteorite*. Such falls, when observed, are incredibly spectacular. The brilliant fireball often breaks up into fragments in mid-flight and produces sonic booms as it ploughs through the air at supersonic speeds. These events are only very rarely witnessed, but pieces of meteorites are quite commonly found over the Earth's entire surface.

Meteorites can be divided into three main classes, according to their compositions. The first are *siderites* – these are roughly 92% iron, 7% nickel and contain traces of cobalt and other minerals. The second class is the *lithosiderites* – containing mineral and metallic elements in various proportions. The third class is the *stony meteorites*. 95% of all the collected meteorites are of the lithosiderite variety.

The study of meteorites is an important field of research because they provide samples of extraterrestrial material thousands of millions of years old. They give direct clues to the composition of the protosolar nebula, and are very much more cheaply obtained than the several billion dollars it cost to procure the relatively small amount of Moon rock.

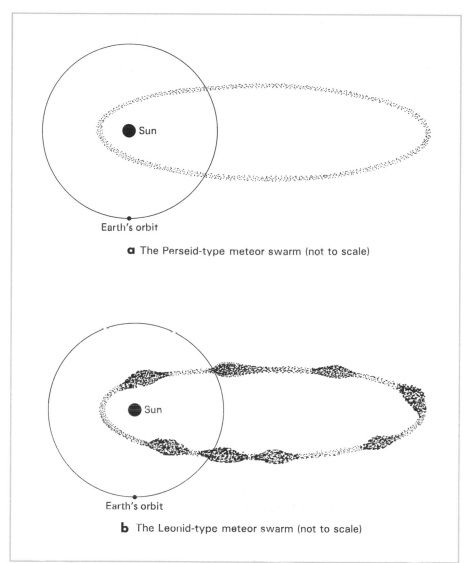

a The Perseid-type meteor swarm (not to scale)

b The Leonid-type meteor swarm (not to scale)

Figure 12.3 Highly idealised (and exaggerated) representation of the difference between **a** a meteor stream that produces a similar shower each year (such as the Perseid) and **b** one that produces a shower that varies from year to year in intensity (such as the Leonid – though in the case of the Leonid swarm there is one main concentration, not several). In practice, the orbits are inclined so that the Earth's and the meteor's orbit cross at only one point.

12.8 The Breakdown of Bode's Law

Before completing our very brief look at the debris of the Solar System we must, equally briefly, return to the question of Bode's law and its validity. It had been exceedingly successful in its accordance with the actual distances of the planets out to the orbit of Uranus, and can even be said to have "predicted" the discovery of a body (or bodies) orbiting at a mean distance of 2.8 AU from the Sun. However, it must be remembered that Bode's law is simply a numerical artefact. As for the prediction of the minor planet swarm, well, the large gap between Mars and Jupiter would suggest to almost anyone that the gap ought to be filled with some body. In fact Kepler, in the early seventeenth century, was so troubled by the odd gap in the planetary orbits that he wrote: "between Mars and Jupiter I put a planet".

The planet beyond Uranus should have an orbital radius of 38.8 AU according to Bode's law. When Neptune was discovered it was found to have an orbital radius of 30.1 AU, thus shattering the law. The discovery of Chiron, orbiting midway between Saturn and Uranus is equally problematical. On the other hand it is of interest to note that the mean distance of the tiny planet Pluto from the Sun is 39.4 AU, not much different from the value predicted for the eighth planet. It should be remembered that Pluto is distinctly odd in that during part of its orbit it approaches the Sun closer than does the planet Neptune. At such times Pluto **is** the eighth planet in order of distance from the Sun.

Where does all this lead us? Some theorists think that the distances of the planets from the Sun can be predicted from a complex theory, perhaps involving the prior establishment of the massive planet Jupiter in its orbit. The various tidal forces and resonant effects might then control the stable orbits in the rest of the Solar System. Perhaps this theory approximates to give a sequence like Bode's Law. However, this is mere conjecture. At present no such theory exists.

Questions

1 Outline the formation and early evolution of the Solar System, explaining the main cratering processes that occurred on the Moon and the terrestrial planets.

2 Write an essay about Bode's law, stating the main cases of observational confirmation and disagreement with the "law".

3 Your Great Aunt Matilda writes you a letter in which she states she saw a bright streak of light, that vaguely reminded her of a Guy Fawkes Night rocket, rapidly shooting across part of the night sky before it fizzled out. She asks you if you know what it could possibly be. She writes that she thinks she has heard, somewhere, of "the Celestial Police" and she wonders if she should contact them to let them know about a possible extraterrestrial invasion! Write her a letter explaining clearly what you think she saw, and dealing with her query concerning "the Celestial Police".

4 Write an essay about meteors, outlining their nature, appearance and paths.

5 Find out about and write a short essay on meteorites. How are meteorites related to asteroids?

6 Who were the "Celestial Police"? What was their purpose and what did they achieve?

Chapter 13

The Comets

THE name "comet" derives from the Greek "aster kometes", which means "long-haired star" and this is a fair, if rough, description of the appearance of a comet. Comets used to be regarded as the portents of doom and tragedy and many a bright comet has caused widespread panic. For instance, a bright comet caused alarm in the England of 1665, in the early weeks of the Great Plague. Another appeared in 1666 just before the Great Fire that was to purge the plague-swept London.

A great comet appears as a condensed fuzzy mass with a magnificent luminous tail stretching across much of the sky. This century such comets have proved to be rare. Several comets are discovered each year, but most of these are relatively insignificant, appearing only as faint and rather fuzzy "blobs", even when seen through powerful telescopes. Indeed, most comets never become bright enough to be visible to the unaided eye. Even so, comets look much more impressive than they really are. In fact, the visible part of the comet is made up of an extremely tenuous mixture of gas and dust that more closely approximates to a vacuum than anything we can produce in a laboratory here on Earth.

13.1 Comet Orbits

Aristotle believed that comets are objects that are ejected by the Earth at tremendous speed and subsequently catch fire in the atmosphere. If comets were really atmospheric phenomena then they might be expected to show the effect of parallax. That is to say, an observer at a particular location might see a comet in a certain position relative to the star background. An observer at another location would see the comet projected against a different position relative to the stars. Tycho Brahe carried out an investigation in 1577 and found no evidence for such parallax effects. He thus concluded that comets were at least as far away as the Moon and probably a lot further.

Tycho's conclusions were slow to be accepted, but Edmond Halley supplied the next advancement when he used mathematical methods, based on Newton's universal law of gravitation, to plot the paths of several bright comets. In the process he found that the comets observed in 1531, 1607 and 1682 were one and the same object – a comet moving round the Sun in an elliptical orbit with a period of 76 years. Halley predicted that the object should return in 1758, which it did, sixteen years after his death. In fact, a detailed analysis showed that the famous comet present for the Norman invasion of England in 1066 was none another than Halley's comet (the name given to it posthumously). Scientists of the time gradually came round to thinking that all comets were members of the Solar System. Halley's comet was last seen in 1986, though not well seen from the northern hemisphere. Its orbit is illustrated in Fig. 13.1.

Comets seem to be fairly neatly divided into two distinctive classes, those which frequently return to perihelion (the *periodic comets*) and those which take centuries, or thousands of years, or perhaps millions of years to return to the inner regions of the Solar System (the *long period comets*). All the long period comets have exceedingly elliptical orbits whose aphelia are at enormous distances from the Sun (of the order of 150 000 AU). Many such comets have been seen only once in recorded history.

Most of the comets we detect are of the *long period* variety. Only about a hundred are known with orbital periods of less than two centuries. Halley's comet is the only bright member of the group of periodic comets, most of the others being rather

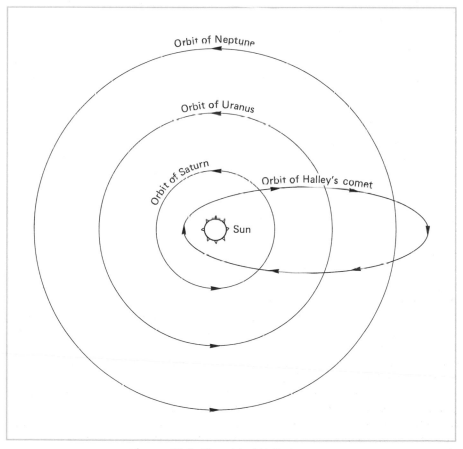

Figure 13.1 The orbit of Halley's comet.

faint telescopic objects. Halley's comet, in common with about half the total number of observed comets, orbits in the wrong-way, or retrograde, direction. *Short period comets* have orbits which are much less elliptical than their long period counterparts. Indeed, some periodic comets have orbits which are very nearly circular.

Comets are very light and insubstantial bodies and their orbits are thus very easily modified by the major planets, particularly Jupiter and Saturn. A number of comets have aphelia about 5 AU from the Sun. The planet Jupiter, orbiting at 5.2 AU, has been responsible for transferring some long period comets into the periodic variety. Encke's comet, which returns to perihelion every 3.3 years, is one example of a "Jovian family" comet.

13.2 The Structure of a Comet

Comets vary appreciably in appearance, but in general the structure is that of a small body, the *nucleus*, surrounded by a distended atmosphere of dust and gas, the *coma*. Often a tail develops from the coma as the comet approaches the Sun, only to shrink again as it recedes into space, having passed perihelion. Not all comets develop tails. The nucleus appears bright because of the concentration of gas and dust around it, but the actual solid nucleus is far too small and dim to be seen. Figure 13.2 shows the appearance of the comet *IRAS–Araki–Alcock* (we name comets after their discoverers) and Fig. 13.3 (*overleaf*) is a representation of the structure of a hypothetical comet. We will now go on to consider the various features of a comet in more detail.

13.3 A Comet's Tail

Comets can have two types of tails and often both occur together. *Type 1* tails are made of ionised gas and *type 2* tails are made of dust. The spectra of type 1 tails reveal ionised molecules and radicals such as: CO^+, CO_2^+, OH^+, CH^+, CN^+, etc. It is only due to the exceedingly tenuous nature of cometary tails that these radicals can remain uncombined. They certainly cannot do so under normal, earthly, conditions. Type 1 tails shine with a bluish colour caused by the ionised gases absorbing short-wave radiations from the Sun and reradiating the energy at visible wavelengths. Type 1 tails always point away from the Sun, deviating no more than a few degrees from the antisolar direction; this means that when a comet has passed perihelion it is travelling tail first (see Fig. 13.4, *overleaf*).

Type 1 tails are usually fairly straight, but bright knots are sometimes seen to form in them and then thread their way along, moving away from the nucleus. The cause of these knots and the direction of the tail lies with the Sun.

The Sun continually sends out a stream of electrified particles, spreading radially into space, the *solar wind*. The particles carry intricately confused magnetic fields with them and these fields interact with the ions in a comet's tail, pushing them away from the Sun. Astronomers have studied photographs of many comets in order to calculate the accelerations of any transient features which form in their tails. In this way they have shown that repulsive forces act on the ions in the

Figure 13.2 The comet *IRAS–Araki–Alcock*. Photograph taken by John Alexander of the Royal Greenwich Observatory on 1983 May 9^d 23^h UT. The telescope (a special camera of 13 cm aperture) was guided on the comet during the 30 minute exposure, so causing the star images to trail. Although not very spectacular to look at, this comet had the distinction of passing within 0.03 AU of the Earth on May 11^d, closer than any other had come since 1770. RGO photograph.

tail which are of the order of a hundred times bigger than the solar gravitation at the distance of the comet.

The exact mechanisms producing these repulsive forces are still not properly understood, but they are thought to involve a two-stage process of magnetic coupling. The idea is that high-energy solar wind particles and the electromagnetic radiation from the Sun cause gaseous material in the coma to become ionised. These ions are then selectively carried away from the coma, in the antisolar direction, by the magnetic field patterns that are "frozen into" the expanding solar wind. When the comet is far away from the Sun it is moving in a roughly radial direction to the centre of the Solar System. When near perihelion, the comet has

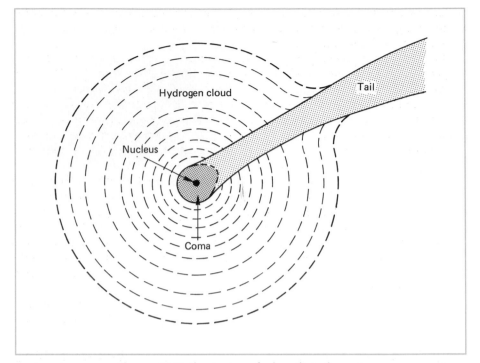

Figure 13.3 The structure of a hypothetical comet.

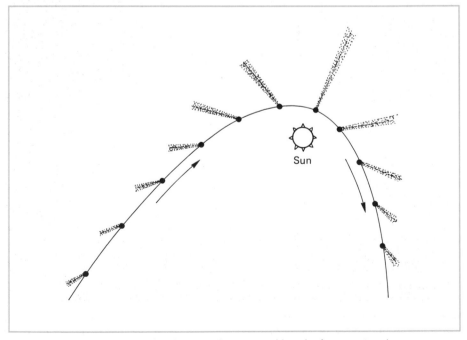

Figure 13.4 The changing direction and length of a comet's tail.

an appreciable transverse motion through the solar wind. This causes the tail to become slightly curved in the same manner as the smoke rising from the funnel of a moving steam locomotive.

Type 2 tails are generally broader and far more curved than type 1 tails. They appear yellowish in colour and have spectra similar to that of the Sun. This indicates that they are composed of colloidal-sized "dust" particles (nothing whatever in common with household dust) which reflect the sunlight. The large amounts of curvature seen in type 2 tails show that, for this type of tail, the repulsive forces supplied by the Sun are only just large enough to overcome the attractive gravitational force. We think that the dust particles are driven outwards by the tiny pressure that electromagnetic radiation exerts on matter.

Some comets develop extensive ion tails but only weak dust tails. For others the reverse is true. Occasionally a comet sports an "anti-tail", an apparently sunward pointing spike, but this is an illusion caused by us seeing part of the highly curved dust tail projected sunward of the Earth–comet line. Some comets have only short and faint tails. Some have none at all. On rare occasions they can develop tails which grow longer than the distance separating the Earth from the Sun. The way that a comet's tail grows as it nears the Sun and wanes as it recedes tells us a lot about the nature of the tail, the coma and the nucleus.

Figure 13.5 shows the comet Hyakutaki (actually C/1996 B2 – a special code given to distinguish one comet from another; many comet hunters, like the

Figure 13.5 Comet Hyakutaki photographed by the author on 1996 March $24^d\ 01^h\ 12^m$ UT, using an ordinary camera mounted piggyback on a telescope. The telescope tracked on the stars for this 1 minute exposure.

Japanese amateur Hyakutaki, discover more than one comet). This was a spectacular sight in a dark sky in the spring of 1996. Note its long, type 1 tail. This was the brightest comet for many years, but at the time of writing (summer 1996) another very bright comet, named Hale–Bopp, is heading towards perihelion and may become a spectacular sight in the spring of 1997.

13.4 The Hydrogen Cloud

A fairly recent (1970) discovery is that vast clouds of hydrogen gas surround the nuclei of comets. Typical cloud radii are of the order of millions of kilometres. The discovery arose when spacecraft observations were made of the comets Tago–Sato–Kosaka and Bennett in a wavelength range inaccessible from the surface of the Earth because of our atmosphere. We think that the clouds are maintained by sunlight splitting the hydroxyl (OH) radicals. The very light hydrogen gas then escapes into space as a vast, expanding, cloud.

13.5 A Comet's Coma

The coma of a comet is a roughly spherical body of gas and dust, centred on the nucleus, which extends to perhaps 100 000 km, or even 1 million km, from it. It shines by means of a combination of reflected sunlight and the fluorescence effects responsible for the visibility of the ion tail. The gaseous component of the coma consists mainly of neutral atoms and molecules, such as: H, OH, NH, NH_2, HCN, CH_3CN, CS, S, O, CO, C, C_2, C_3 and even traces of metals such as sodium and iron. As is the case for a comet's tail, it possesses no coma until it approaches the inner regions of the Solar System. We now realise that the nucleus is the storehouse of materials for the coma and the tail. They are only released from deep-freeze when the nucleus nears the Sun's heat.

13.6 A Comet's Nucleus

Although a comet may possess an enormous coma, surrounded by an even larger hydrogen cloud and a tail that might be longer than an astronomical unit, by far the greater part of the comet's actual mass lies in a tiny body near the centre of the coma – the nucleus. This is a sort of "dirty snowball", a few kilometres across, composed of various types of ices. Perhaps many comets have a small core of rocky material at the centre of their nuclei, but we cannot be sure at present. As the nucleus of the comet approaches the Sun, so it is heated and volatiles are released from its surface to form the coma and the tail. The gas release tends to occur in jets and this causes significant perturbations of the comet's path as it wings its way through the inner Solar System.

The fact that many comets have been observed to pass through their perihelia dozens of times (Halley's comet has been reliably documented as far back as

240 BC) means that they must possess large reservoirs of volatiles in "cold storage", as it were. However, not all comets survive close passages past the Sun. There is also the question of what, if anything, is left when a comet has eventually exhausted its supply of volatiles. These matters are briefly discussed in the next section.

13.7 Cometary Debris

Despite the terror that they caused at one time, comets are very harmless things. The Earth has even passed through the tails of several comets without any ill effects. However, a direct collision with a comet's nucleus might be a different matter. In Russia, on 30 June 1908, a brilliant fireball was seen to streak across the morning sky. It terminated in a resounding explosion that was heard over a thousand kilometres away. All over the world seismometers recorded the shock-waves and peculiar atmospheric effects followed.

Investigators later found that a large area of forest in the Tunguska region of Siberia (the estimated explosion site) was decimated. The trees were knocked flat and stripped of foliage for many kilometres around the central region. No crater was found, so no ordinary meteorite had caused the explosion. Opinions vary, but it is quite likely, though by no means proven, that a cometary nucleus entered the Earth's atmosphere and produced the observed effects. This at least is by far the likeliest explanation; and although opinions vary, we are probably quite right to disregard the proposals of extraterrestrial spacecraft and anti-matter missiles that have come from some quarters!

Astronomers had a marvellous chance to observe the effects of a cometary impact on a planet (fortunately not our own!) when the fragments of the disintegrated Shoemaker–Levy 9 rained down on the planet Jupiter. The effects on the Jovian planet were certainly spectacular, as is described in Chapter 9.

Most comets survive perihelic passage intact but not all – a famous instance of the breakup of a comet is that of the one discovered by Biela in 1826. It was found to be a member of the Jovian family and possessed an orbital period of just under seven years. The comet was well observed when it next returned in 1832 but was missed in 1839 because of unfavourable conditions. It was next seen in 1845. However, the comet astounded astronomers by splitting into two as it passed perihelion. The two parts were roughly equal in brightness but one had a tail and the other did not. The next return, in 1852, once again heralded the two parts of the comet, though they were further apart. Conditions were very unfavourable for the 1859 return to perihelion but the comet(s) should have been well seen again in 1866 – it was not! The comet also failed to show in 1872. However, a rich shower of meteors occurred on the date when the Earth passed through the orbital path calculated for the comet.

Since then, every time the Earth has transgressed the path of the old comet, so a shower of meteors has been observed. Over the years the shower has lost its liveliness, but a few "comet" meteors may still be seen on 27 November every year. We can be fairly sure that Biela's comet was tidally disrupted by the Sun and that a shoal of particles has now spread along its old orbit. This incident confirmed what was already suspected – namely that meteor showers and comets have some connection. For instance the Aquarids, occurring in May each year, are associated with the orbit of Halley's comet.

The case of Shoemaker–Levy 9, ripped asunder as it strayed too close to Jupiter, has already been mentioned but even without the nucleus neccessarily being disrupted, the expanding comas and streaming tails of comets must shed material into the inner regions of the Solar System.

Comets can lose appreciable mass at each perihelic passage and short period comets are, by astronomical standards, fairly short-lived objects. Some of the cometary debris is seen from the Earth as the *zodiacal light* and the associated *gegenschein*, also known as the *counterglow*. The zodiacal light appears as a faint band of radiance extending from the horizon, where it is thickest and most apparent, upwards along the zodiac. It forms a complete arch, extending from horizon to horizon. A local thickening and brightening occurs at a point in the sky diametrically opposite to the Sun. This is the gegenschein. You need to observe in a very dark sky, well away from light pollution, if you are to see the zodiacal light. The gegenschein is even more elusive and most observers will never see it.

We see the fine cometary dust that forms the zodiacal light because it back-scatters sunlight. In order to maintain this zodiacal illumination, comets are estimated to contribute about 10 tonnes of dust per second to the inner Solar System. Over a period of years the dust is blown away by the solar radiation, only to be replaced by that from more comets passing near the Sun.

Do comets have rocky cores at the centre of their nuclei? If they do then this core must survive after the volatiles have all been evaporated by successive passages past perihelion. Perhaps several of the observed asteroids are the remnants of old periodic comets. At present astronomers do not know the answers to these questions. This is why comets are so eagerly studied when they make their appearances before withdrawing once again into the darkness and coldness of space from whence they came.

13.8 Spacecraft to Halley's Comet

The 1986 return of Halley's comet was significant in that it allowed, for the first time, the exciting prospect of a spacecraft flying past it at close range. In fact six spacecraft were dispatched to rendezvous with the celestial visitor. The Russians launched *Vegas 1* and *2* in December 1984. Both these spacecraft passed the nucleus, at a distance of 8000 km, on 6 March and 9 March 1986, respectively. Next launched was the Japanese probe *Sakigake*, in January 1985. It passed within 4 million km of the comet's nucleus on 11 March 1986.

Then came the European Space Agency's probe *Giotto*, launched in July 1985 and arriving at the comet on 14 March 1986. A second Japanese probe, *Suisei*, was sent on its way in August 1985, approaching within about 200 000 km of the nucleus on 8 March 1986. Finally there was the extremely versatile *ICE* (International Cometary Explorer), launched in December 1978 to study the solar wind. NASA controllers used the gravitational influence of the Moon to direct this probe through the comet Giacobini–Zinner, in September 1985, then on to Halley's comet, which it passed on 28 March 1986. This probe stayed well away from the nucleus of the comet (nearly 40 million km from it) but investigated how the solar wind interacted with the cometary material. Between them, the spacecraft have gathered much information on the nature of Halley's comet as well as how it interacts with the material in interplanetary space.

The space probe *Giotto* is of special interest, since it passed through the comet's inner coma, passing the nucleus at a distance of 500 km. It took colour pictures as well as making many physical measurements of the gas, dust and plasma associated with the comet. As the probe ploughed through the coma, its velocity relative to the dust particles was of the order of 70 km/s. The project organisers had always feared that the probe might not survive the encounter because of dust particles striking it at high speed. It was fitted with a double shield designed to withstand the hits of all but the largest particles that were expected. In the event, the spacecraft did survive but seconds before closest approach it was hit by a larger-than-average particle which caused the probe to be jarred enough to tip its antenna out of alignment with the Earth. All data was then lost until communications were re-established after about an hour.

Nevertheless, close-range pictures of Halley's nucleus were obtained and these showed it to be a very dark and irregularly shaped body, about 5 km across. Bright jets of gas and dust issuing from the nucleus are visible in the photographs. Gravimetric measurements indicate that the overall density of the nucleus is about 500 kg/m^3. This is about half the density of most types of ice and is a surprising result. The nucleus must be a very loose agglomeration of ice crystals (of various chemical compositions) and any rocky core must be very small. The dark covering was largely unexpected and is still something of a puzzle. March 1986 was certainly a month to go down in astronomical history, and it is only fitting that the events involved the most famous of all comets.

13.9 The Origins of the Comets

Currently it is thought that the comets reside in a huge, shell-like cloud, called the *Oort Cloud*, after the late Jan Oort, extending over a light year from the centre of the Solar System. Passing stars cause some of the orbiting ice-balls (estimated to be 200 000 million in number – a total mass equivalent to about one-tenth that of the Earth) to fall inwards towards the Solar System, under the influence of the Sun's gravity. Sometimes the major planets further perturb their paths and so cause them to settle into short-period orbits.

It seems likely that the comets were formed from fragmentary interstellar clouds along with the rest of the Solar System. Certainly the object that was first taken to be an asteroid, Chiron (see the Chapter 12), orbiting between Saturn and Uranus, now seems really to be an unusually large comet nucleus. However, it must be remembered that scientists ideas are still very theoretical. The comets remain as mysterious as they are beautiful.

Questions

1 Describe the appearance, nature and structure of a typical comet.

2 Describe the path of a typical comet through the Solar System. What are the reasons for the fact that a comet only spends a very short time at perihelion?

3 Distinguish between *long-period* comets and *periodic* comets. What is the cause of the difference between them?

4 Describe how, and why, the appearance of a comet changes as it approaches the Sun and later recedes back into space.

5 Write a detailed essay on the subject of the tails of comets.

6 Write a short essay about the *hydrogen cloud*, *coma*, and *nucleus* of a comet.

7 Write a short account of the relationship between meteors and comets, as well as other cometary debris.

8 Briefly outline our ideas about the possible origin of comets, explaining how they eventually arrive in the inner regions of the Solar System.

Chapter 14

The Sun

E VEN if the masses of all the planets, satellites and other bodies of the Solar
System could be added together, they would form a body only about a thousandth of the mass of the Sun. As well as being our planetary system's dominant
body, the Sun is also the sole source of the copious amounts of light and heat so
necessary for life on the Earth. The Sun is truly the king of the Solar System.

14.1 The Size and Composition of the Sun

When seen through a thick mist the Sun appears as a luminous disk that subtends an apparent angular diameter of slightly greater than half a degree. Our
Earth orbits the Sun at a mean distance of 150 million km and so it is a simple
matter to work out its diameter and volume. It turns out to be a vast globe
1.4 million km in diameter (about 109 times the diameter of our Earth). It thus
has a volume equivalent to more than a million Earths. The values of the Earth's
mass (known from laboratory experiments) and its orbital distance and period
(known from astronomical observations) allow us to use Newton's law of gravitation to calculate the mass of the Sun. It turns out to have the incredible value of
2×10^{30} kg, i.e. 330 000 times that of the Earth. From its mass and volume, the
mean density of the Sun works out to be 1410 kg/m^3. This is about 1.4 times as
dense as water.

However, the globe of the Sun is not of uniform density throughout but is very
rarefied at its outer edge and very dense towards its centre. The values of density
range from 1×10^{-6} kg/m^3 (one-millionth of a kilogram per cubic metre) to about
1.5×10^5 kg/m^3 (150 000 kilograms per cubic metre). By comparison, the density
of the atmosphere at the surface of the Earth is 1 kg/m^3, making the visible
"surface" of the Sun very rarefied indeed.

The light from the Sun can be analysed by means of a spectroscope, and so it is
possible to find out what it is made of, even if a probe cannot be sent there to
scoop up a shovelful of the solar material. Such analysis leads us to believe that
the Sun is largely composed of hydrogen (about 73% by mass), with helium as the

next major constituent (25%), followed by trace amounts of the rest of the elements. It is thought that the Sun has a fairly uniform chemical composition throughout its globe.

14.2 The Absolute Scale of Temperature

The Fahrenheit and Centigrade (Celsius) scales of temperature are fine for everyday use but they have rather arbitrary foundations. The Fahrenheit scale has 180 divisions or (degrees) separating the temperatures of boiling and freezing water (under normal atmospheric pressure), with the freezing point being fixed at 32 °F. In the Centigrade (Celsius) scale 100 degrees separate the boiling and freezing points of water with the freezing point being fixed at 0 °C.

Astrophysicists prefer to use a scale that has a more fundamental foundation in thermodynamic theory. This is the *absolute*, or *Kelvin*, scale of temperature. The zero point of this scale ("absolute zero") is set at the temperature at which all molecular motions cease (at any higher temperature the molecules in a gas are in constant random motion – this gives rise to gas pressure, because of the molecules striking the walls of the containing vessel). On the Centigrade scale absolute zero is –273 °C. For convenience scientists have set the division (degree) of the absolute temperature scale equal in size to the divisions of the Centigrade scale. This allows for easy conversion. A temperature on the absolute scale is measured in kelvins (K), **not** degrees Kelvin. To convert temperatures from the Centigrade scale to the absolute scale all one has to do is add 273 (see Table 14.1).

14.3 The Structure of the Sun

The temperature of the Sun's visible surface, or *photosphere,* averages 5800 K, but this increases rapidly to an estimated 15.6 million K at the Sun's centre, while the central pressure is calculated to be a colossal 250 thousand million Earth atmospheres. Under these extreme conditions nuclear reactions occur which release energy and keep the Sun shining.

Nobody has actually seen inside the Sun, but astrophysicists can extrapolate the information we get from studying its surface, as well as using the fundamental

Table 14.1 Temperature conversion table

Absolute temperature (in K = kelvins)	Centigrade temperature (°C)
0	–273
173	–100
273	0
373	100

laws of physics and applying our understanding of the behaviour of matter, to deduce the nature and conditions within the solar globe.

Recently this has been added to by the study of *helioseismology*. Seismologists have built up a picture of the Earth's interior by studying the vibrations generated by earthquakes. The Sun also vibrates very slightly in a number of different ways and with a range of frequencies. These can be analysed by studying the slight shifts in the frequency of the Sun's light they produce on its surface (this is an application of the Doppler effect, discussed in Chapter 15). Helioseismologists make use of this data to infer the structure of the Sun's interior regions. The Global Oscillation Network Group (GONG) has just (1996) come into operation to intensively study the Sun by means of observing its oscillations.

Also, a dedicated spacecraft, *SOHO*, standing for 'Solar and Heliospheric Observatory', was launched at the end of 1995 and was parked in a position of neutral gravity (a *Langrangian point*) about a hundredth of the way along a line from the Earth to the Sun. From that vantage point it can continually monitor the Sun and the solar wind from outside the Earth's magnetosphere. As well as observing the Sun in a variety of different wavebands, it is also equipped to do helioseismology. *SOHO* is expected to make and transmit observations for at least $2\frac{1}{2}$ years. It is already providing valuable clues to the answers to many vexing questions.

The weakest link in the chain of reasoning is our present understanding of the detailed behaviour of matter under the extreme and unfamiliar conditions within the Sun. Nevertheless, a picture of its interior is emerging which is unlikely to be very wrong in its major details. Figure 14.1 (*overleaf*) represents pictorially the structure of the Sun as we understand it at present.

Most of the Sun's supply of energy is created in its central core. The core emits most of this energy in the form of short-wave electromagnetic radiations, namely X-ray and γ-ray photons (see the next chapter). These photons do not travel very far (only about 1 cm) before being absorbed by the material in the solar mantle. However, the matter re-emits the radiation and so it gradually works its way to the surface is what is termed a "random walk". In the process, the radiation gradually loses energy and it takes, perhaps, 170 000 years for it to arrive at the photosphere. By the time it does so, most of the energy is in the form of visible light.

The energy-producing core of the Sun extends from the core to a quarter of its total radius. The process of absorption and re-emission is the dominant mode of energy transport out to 71% of the Sun's radius. The region between the core and this radius is known as the *radiative zone*. Outwards beyond the radiative zone the temperature is low enough to allow electrons to be captured by nuclei to form partially ionised atoms (below this level the gas is totally ionised). The partially ionised atoms absorb photons very much more readily than the material of the radiative zone. As a result, a steep temperature gradient develops within the gas and this causes convection. Thus between the radiative zone and the photosphere we have the *convective zone*, where convection takes over as the major process of energy transport.

The top of the photosphere marks the edge of the main body of the Sun, though layers of very low density gas extend through the Solar System and merge into interstellar space. A layer of gas, called the *chromosphere*, lies above the photosphere. The features known as *spicules* and *prominences* extend from the

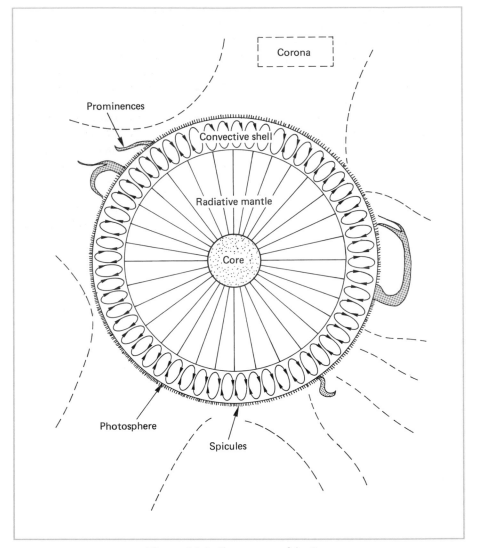

Figure 14.1 The structure of the Sun.

chromosphere. Moving out from the chromosphere we have the *corona*, which is effectively the Sun's atmosphere. Each of these features will be discussed in detail in the following sections.

14.4 Solar Eclipses

Chapter 6 explains how the Moon could sometimes pass into the shadow cone in space formed by the Earth, so creating a lunar eclipse. Of course, it is also possible for the Moon to pass between the Sun and the Earth. We then see a solar eclipse.

Figure 14.2(a) is a very out-of-scale representation of how a solar eclipse occurs. (It is a remarkable coincidence that the Sun and the Moon both appear the same apparent size as viewed from the Earth; the ratio of the diameter of the Moon to its mean distance from us just happens to be the same as the ratio of the Sun's diameter to its mean distance.) It should be obvious that a solar eclipse can only occur at the time of the new Moon and when the Moon is at the ascending node or the descending node in its orbit.

Solar eclipses can be both *partial*, where the Moon appears not to cover the entire solar disk, and *total*, where the Sun's disk is completely covered. As can be seen from the diagram, a total solar eclipse can be seen only from a restricted region of the Earth's surface at any one moment during the eclipse's progress. All other regions see, at best, a partial eclipse. In fact, due to the rotation of the Earth as well as the relative motions of the Sun and the Moon, a narrow track is swept across the Earth's surface in which the eclipse can appear as total. Naturally, from any given location the eclipse can only have a limited duration, never exceeding eight minutes.

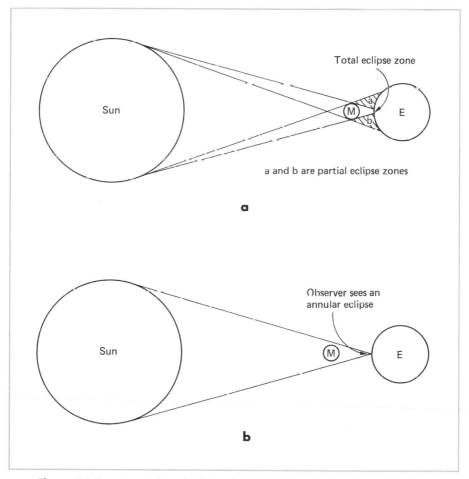

Figure 14.2 a A partial/total eclipse of the Sun. **b** An annular eclipse of the Sun.

However, the duration of totality is usually much smaller than this, and it varies from eclipse to eclipse. This is because the Earth and the Moon both move in elliptical orbits and so the apparent sizes of the Sun and the Moon both vary. A solar eclipse will have the longest possible duration if it occurs at a time when the Earth is at aphelion and the Moon is at perigee. In the converse situation the apparent diameter of the Sun would be larger than that of the Moon. In that case an *annular eclipse* would be formed. This is so called because the Moon does not entirely cover the Sun but leaves, at mid-eclipse, a bright ring of sunlight surrounding the disk of the Moon. Figure 14.2(b) shows the principle of the annular eclipse.

A total solar eclipse is an even more impressive spectacle than is a total lunar eclipse. The partial phase lasts about an hour, with progressively more and more of the Sun being covered by the Moon until only a thin crescent remains. The sky then rapidly darkens and the last bright chink of sunlight vanishes. At this point the sky becomes dark enough to see the brightest stars. From around the Moon's limb springs the pearly white glow of the solar corona, and little red tongues of "flame" mark the solar prominences. After a brief few minutes of this spectacle a sliver of bright sunlight suddenly explodes into view as the rays from the Sun find their way through some depression in the lunar limb to form the famous "diamond ring" effect. Totality is over, and during the course of the next hour more and more of the solar disk is uncovered and the eclipse becomes only a memory.

14.5 The Photosphere

On no account should any ordinary telescope, or a pair of binoculars, be turned towards the Sun. The brilliant light and the intense heat would instantly cause irreparable damage to the eye of the observer. One method of observation that is safe is to use the telescope to project the solar image onto a card held behind the eyepiece. If the card is shaded from direct sunlight then a bright image of the Sun can be obtained that is at least twice as large as the aperture of the telescope. If the disk is enlarged to 15 cm or so then one feature will be very apparent – it is not as bright close to its edge as it is near its centre. This effect is called *limb darkening* and is caused by the Sun's gaseous nature.

As well as emitting light because of its high temperature, the photosphere also absorbs light. The temperature, density, and thus the brightness of the photosphere, increases with depth. When we look at the centre of the solar disk we can see down to a depth of a few hundred kilometres before the photospheric material becomes totally opaque. When we look near the limb we are effectively looking along a greater path-length of solar material and so cannot see to as great a depth below the top of the photosphere before the radiation is totally absorbed. Figure 14.3 shows the principle. The higher layers in the photosphere are not as hot and bright as the lower layers, and hence the darkening towards the limb. The photograph of the Sun (Fig. 14.4) shows the limb darkening very well.

The best Earth-based photographs show that the photosphere is not a smooth and uniform layer but is made up of a seething mass of turbulent cells, which give the Sun a grainy appearance. These bright *granules* are each about 1000 km across and are the tops of turbulent columns of material rising from the lower, hotter, layers. The solar material cools and mushrooms outwards, then falls between the rising columns. The average lifetime of a granule is around ten

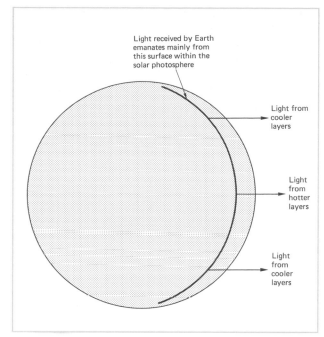

Light received by Earth emanates mainly from this surface within the solar photosphere

Light from cooler layers

Light from hotter layers

Light from cooler layers

Figure 14.3 Explanation of limb darkening (see text).

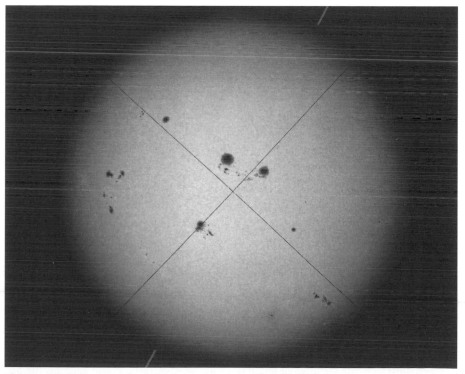

Figure 14.4 Photograph of the Sun taken using the photoheliograph at Herstmonceux on 1970 April 9[d]. RGO photograph.

minutes. Larger-scaled convective motions produce *supergranules*, which are rather similar to granules but on average about two hundred times larger. Conventional photographs do not show up the supergranular cells, special techniques being needed to reveal them. Other large features are often to be seen on the Sun, namely the *sunspots* and the *faculae*, transient features intrinsically involved with the Sun's powerful magnetic field.

14.6 The Solar Magnetosphere

The Sun does not rotate on its axis in the same manner as a solid body. Its equatorial regions take 25 days to go once round, but this period increases to about 35 days near the poles. Since the Sun is almost entirely composed of electrified gas, or *plasma*, any motion of its material gives rise to enormous electrical currents (an electrical current is simply the movement of electrical charges). Any electrical current creates a magnetic field. The bigger the current, the bigger the magnetic field strength.

Because of the differential rotation of the Sun and the turbulent nature of its outer layers, its magnetic field is extremely powerful and very complicated, as you might expect. The differential rotation tends to "wind up" the general north–south field into "flux tubes" which approximately wrap around the Sun, though breaking out of the surface in complicated loops and twists. The seething outer layers of the Sun also add to the confusion of the magnetic field. The value of magnetic field strength in the polar regions is usually around 1×10^{-4} T (teslas), roughly twice the field strength at the surface of the Earth. Elsewhere the magnetic field is very much stronger, reaching a value of over 0.3 T in the regions of the sunspots. To complicate matters even further, the north–south polarity of the Sun reverses over a cycle of around 22 years.

14.7 The Sun's Internal Rotation

One of the biggest surprises yet to emerge from helioseismology concerns the way the Sun rotates at different depths. It seems that the differential rotation so obvious at the photosphere extends right down to the base of the convection zone. Below this the equatorial rotation slows, while the higher-latitude rotation period increases. At about halfway from the photosphere to the core the two rotation periods become the same. The first indications are that below this the Sun rotates as one solid body. However, at the time of writing further study is required, as the results concerning these depths are far from being unambiguous. Certainly, though, nobody expected the Sun to rotate in this way.

14.8 Sunspots and Faculae

The earliest records we have of small dark areas, sunspots, being seen on the Sun go back over 2000 years to the civilisations in ancient China. The observations were, of course, of the spots that were large enough to be seen with the naked eye when the disk was suitably dimmed by haze. Galileo first applied his telescope to solar observation in late 1610 and he saw many smaller sunspots.

A typical sunspot consists of an irregularly shaped dark area, the *umbra*, surrounded by a lighter area, the *penumbra*. The umbra and penumbra are both darker than the photosphere because they have lower temperatures. Remembering that the average photospheric temperature is 5800 K, the penumbra of a typical spot ranges around 5500 K, while the umbra is usually not much hotter than 4000 K. Thus a sunspot appears dark only by contrast; if it could be seen shining by itself it would appear very bright indeed.

Sunspots are carried round by the rotation of the Sun. They form, develop and change, and finally die away over a period of time. The larger spots survive for two or three months. The number of sunspots on the solar disk varies over a cycle of approximately eleven years. The latitudes where most of the spots occur also changes over the cycle. At the beginning of an eleven year period sunspots mostly occur at latitudes of 35° north and south. By the end of the period they mostly occur near the equator. At the same time the next cycle begins with a few high-latitude spots. The incidence of sunspots can be represented on a diagram with their latitudes. Figure 14.5 is an illustration of such a diagram, known as a "butterfly diagram" for obvious reasons.

Sunspots form where magnetic field lines loop in and out of the solar surface. They thus tend to form in pairs of opposite magnetic polarity. In some way the magnetic field lines carry away energy from the region where they intersect the photosphere, so leading to a lower temperature – and hence the visible sunspot. Exactly how this mechanism works is not yet known.

The fact that the sunspot cycle is half the period for the magnetic reversal of the Sun is not thought to be coincidence. Figure 14.6 (*overleaf*) shows the general idea. At the beginning of a cycle the field might be as shown in the highly idealised diagram in Fig. 14.6(a). As the cycle progresses, so the solar magnetic field gets wrapped around the Sun, forming concentrated flux tubes as in Fig. 14.6(b). These flux tubes then become kinked and buoyant in the overall field, so they

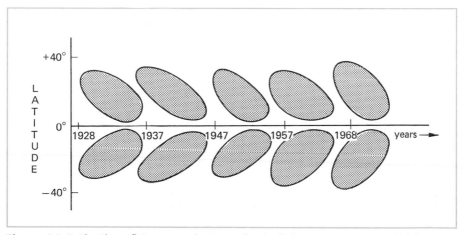

Figure 14.5 The "butterfly" sunspot diagram. The shaded areas are composed of the times and latitudes of the recorded sunspots.

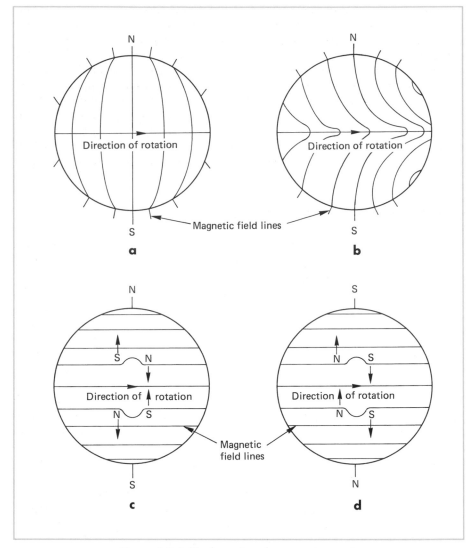

Figure 14.6 The formation of sunspots (see text).

break out of the Sun's surface. The leading member of the pair of spots will have the same polarity as the overall field in that hemisphere, the following spot having the opposite polarity (Fig. 14.6(c)).

The following spot is then attracted in the direction of the pole, while the leading spot heads towards the equator. As the two spots separate, so their linking magnetic fields weaken and the spots eventually fade away. The spots that migrate towards the poles (though they never get beyond latitudes 45° north and south) eventually cancel the polar magnetic field and then change its polarity. Differential rotation then causes the overall magnetic field to unwind itself (the direction of the magnetic field lines is now reversed). When the lines are straight they then start to wrap in the opposite sense, so starting the next 11 year cycle with the north–south field and the sunspots all of opposite polarity (Fig. 14.6(d)).

Figure 14.4 shows several large sunspots. If you look at the photograph very carefully you should see some small, bright, cloud-like features dotted over the surface of the Sun. They are mainly associated with the sunspots but they tend to show up better against the darker limb. These are *faculae*, small regions of higher temperature than the average for the photosphere. Faculae are often associated with clouds of hot material higher up in the chromosphere, termed *plages*. Faculae and plages appear to be caused by local intensifications of the magnetic field and they are certainly areas of concentration of energy. The precise details of the mechanisms responsible for creating them have so far not been discovered.

14.9 The Chromosphere

The chromosphere is effectively the "lower atmosphere" of the Sun. It is a low-density layer of gas, of the same composition as the rest of the Sun, on top of the photosphere. Its temperature varies from about 4000 K just above the photosphere to about half a million K over its thickness of a few thousand kilometres. Its density drops dramatically with height, ranging from 1×10^{-5} to 1×10^{-10} kg/m^3. Thus the top of the chromosphere, though at a very high temperature, has little heat energy associated with it. As is the case for the supergranules, special techniques are needed to study the chromosphere. Its outer edges can be momentarily revealed in total eclipses. It then appears as a thin pinkish band showing round the limb of the Moon. However an instrument called a *spectrohelioscope* allows the Sun to be studied in the light of just one wavelength (which we call *monochromatic light*), and if we select the correct wavelength (for instance that characteristic of the chief emission of hydrogen gas, Hα) the chromosphere can be rendered visible. The same effect can be produced by means of a specially constructed filter. Figure 14.7 (*overleaf*) is a photograph of the Sun obtained through such a filter.

Far from being a homogeneous layer, the chromosphere exhibits detailed structure and seething activity. The most obvious structures are the *spicules*. These are great jets of gas, rooted at the foot of the chromosphere. They rather resemble tall blades of grass sticking up on an otherwise threadbare lawn. A typical spicule consists of gas, at a temperature of about 15 000 K, rising upwards at 25 km/s to a height of 10 000 km. Any one spicule is a short-lived affair. A few minutes after its formation it disperses into the general chromosphere. The spicules are not evenly spread over the solar surface but are arranged around the edges of cells, rather in the same way that the metal in a chain-link fence surrounds the holes. Figure 14.8 (*overleaf*) illustrates this. These cells, forming the so-called *chromospheric network*, coincide with the photospheric supergranules.

Detailed studies show that the spicules follow the local, chaotic, magnetic field lines. When near sunspots the spicules become stretched out and more or less horizontal to the surface. They are then termed *fibrils*. To gain an impression of what a fibril looks like, sprinkle some iron filings on a piece of card and bring up a magnet to the card's underside.

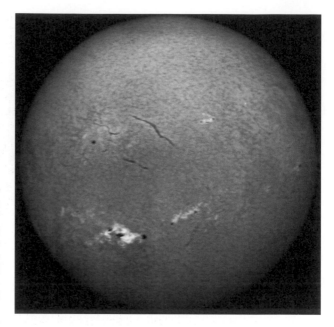

Figure 14.7 The Sun photographed in Hα light on 1981 October 12ᵈ 12ʰ 00ᵐ UT by Eric Strach, using special equipment. Notice the solar granulation visible at this wavelength, as well as the bright faculae surrounding the sunspots and the dark filaments.

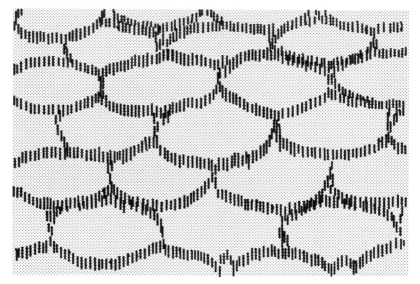

Figure 14.8 The arrangement of spicules on the Sun's surface.

14.10 Prominences and Filaments

Figure 14.9 is a photograph of the totally eclipsed Sun. On it you will see several luminous protrusions, looking like tongues and arches of flame, extending from behind the lunar disk. These are known as *prominences*. They glow with the

Figure 14.9 The totally eclipsed Sun photographed at Sobral, in Brazil, on 1919 May 29[d]. RGO photograph.

characteristic red colour of hydrogen gas and are readily seen when total solar eclipses occur. Special equipment designed to view the Sun in the light of hydrogen, whilst blocking off the direct light from the disk, allows prominences to be studied at other times. If the solar disk is imaged in hydrogen light then the prominences projecting from its edge are not seen (their light is swamped by that from the disk) but dark silhouettes of prominences are seen on it. Some can be seen in Fig. 14.7. These are called *filaments*, though they are exactly the same phenomenon as the prominences. We just see them in a different way.

Prominences (and filaments) are of two forms: quiescent and active. *Quiescent prominences* can survive with little structural change for as long as a year, though a few months is the norm. They often give the impression of upward surges of gas from the chromosphere, though they are really gas condensing from the more tenuous corona. In most cases the gas streams downwards towards the Sun, though upward gas motions are not unknown. The Sun's complex magnetic field seems to support the condensed material, the downward flows of matter mostly following the field lines. They can show a variety of forms – loops, arches, complex filaments, etc., often extending to tens of thousands of metres in height. When they disappear they often do so with a violent eruption which spatters much of the material into space.

Active prominences are much shorter-lived than the quiescent variety, and most of them are much smaller. They can change their forms rapidly, even over a period of just a few minutes. Some active prominences consist of falling material condensed from the solar corona, but many are tremendous upward surges of material. In a few of the most violent events thousands of millions of tonnes of chromospheric material can be thrown upwards at greater than the Sun's escape velocity of 618 km/s.

14.11 The Solar Corona

Visible as a pearly white glow at the time of a total solar eclipse, the *corona* is effectively the Sun's outer atmosphere. If it could be seen shining by itself, the corona would provide nearly as much illumination as the full Moon. However the brilliance of the solar photosphere means that, aside from the times of eclipses, special equipment must be used to study it.

The corona has the same composition as the rest of the Sun and extends to several times the Sun's diameter. Its average density is around 1×10^{-14} kg/m^3. The original measurements made of the radiations produced by the corona, in order to determine its temperature, produced the astonishing value of the order of a million kelvins. The scientists making them had expected the corona to be cooler than the photosphere, not a thousand times hotter! In fact, modern measurements indicate that the temperatures range about 1 million K to 5 million K in different regions. The corona's high temperature is yet another mystery that needs solving.

The corona displays considerable, and variable, structure. During times of minimum solar activity (i.e. when there are few sunspots visible) the corona is fairly regular in shape, but at times of maximum activity it becomes very irregular, with great "streamers" stretching into space above the prominences. Ultraviolet and X-ray photographs of the corona often show dark patches, or *coronal holes*, in the otherwise mottled and translucent corona. These are cooler patches within the corona caused by the Sun's magnetic field being locally weaker and the field lines flowing out into space. Elsewhere the field lines form closed loops and arches with the Sun's surface. Magnetically energised *coronal mass ejections* sometimes occur, where billions of tonnes of plasma are shot into space.

14.12 Solar Flares and the Solar Wind

The Earth and the other planets are bathed in a breeze of electrified particles sent out from the Sun – the *solar wind*. At the orbit of the Earth the particle density is of the order of 2 per 100 000 cubic metres. The velocities of the particles average 500 km/s, though the density and speed of them vary greatly. In fact, the solar wind is very gusty, especially at times of maximum solar activity. The particles are mainly electrons and protons, though heavier ions and nuclei are also present.

Near the Sun, where the magnetic field is strong, the solar wind streams radially outwards from it, co-revolving with the Sun. However, at greater distances the magnetic field is weaker and the particles no longer co-revolve. They form a spiral pattern as they move away from the Sun in the same manner as do water droplets from a lawn sprinkler. Since the solar wind is electrically conductive it is able to trap a magnetic field within it. Those solar magnetic field lines which are open-ended and leave the Sun are "frozen into" the solar wind and take on its spiral pattern. As the particles sweep past the Earth, some can get trapped in the Van Allen zones. These then overspill into our atmosphere, producing the aurorae described in Chapter 2.

The three-dimensional structure and properties of the solar wind are being studied by the spacecraft *Ulysses*, which has been put into a polar orbit around the Sun, as well as its equatorial plane properties by *SOHO*.

As far as we can tell, the solar wind is mainly fed from the expanding corona, though particles must also be supplied from spicules and prominences. The solar wind escaping through coronal holes is particularly strong, and nearly twice as fast as elsewhere, because the particles can then ride the magnetic field lines out into the planetary system. One phenomenon that can give rise to enormous gusts of highly energetic particles is the *solar flare*.

The English astronomer Richard Carrington was the first to see and record a solar flare in 1859. It showed up as a bright patch visible on the photosphere when the Sun was projected onto a screen using a telescope. Such "white light" flares are only very rarely observed, though less energetic events are more often seen in hydrogen light. They are violent eruptions on the photosphere where the temperature in a small region may rise to several thousands of degrees higher than that of the surrounding photosphere. The power involved in a flare can be as high as 1×10^{17} megawatts (MW), nearly a thousandth of the Sun's power output! The major event usually lasts for around twenty minutes and during this time collosal numbers of atomic particles are sprayed into space.

Flares can cause intense auroral activity, magnetic storms and even radio blackouts here on Earth. Flares also pose a real hazard to astronauts in flight because they could be bathed in lethal doses of radiation when the particles reach Earth's vicinity, two or three days after the event. Solar flares are undoubtedly caused by releases of energy from the Sun's magnetic field, possibly by the shearing or reconnection of magnetic field lines, though we are not entirely certain of the mechanism.

14.13 The Sun's Energy Source

That the Sun sends us copious amounts of light and heat is obvious to anyone. What is not so obvious is **how** the Sun produces such a colossal outpouring of energy. Whatever mechanism is considered must also explain the Sun's great longevity. We know that it has been shining for at least 4600 million years with at least approximately its present output.

The value of the energy per second (power) intercepted per square metre by the Earth above its atmosphere (this is always the value quoted to avoid having to allow for atmospheric absorption factors) is termed the *solar constant*. Its value is 1.368 kW/m². Knowing the radius of the Earth's orbit and the radius of the Sun, it is possible to use the solar constant to calculate the power emitted per square metre from the solar surface. This turns out to be 6.3×10^4 kW/m². This means that the total outpouring of energy from the Sun happens at the rate of nearly 4×10^{20} MW!

Until well into this century we simply did not know how the Sun managed to produce so much energy over such a long period of time. We now know that nuclear fusion is responsible. Basically, the extreme conditions of temperature and pressure in the core of the Sun cause hydrogen nuclei to combine together to form helium nuclei. A hydrogen nucleus is just a single proton, but a helium nucleus is made up of two protons and two neutrons (for the type of helium known as helium-4). By a slightly complicated route (see Fig. 14.10), four hydrogen nuclei are combined to form one helium nucleus.

Now, the mass of one helium nucleus is slightly less than four times the mass of a hydrogen nucleus. What happens to the missing mass? The answer is that it is converted into energy. Einstein theoretically demonstrated that mass and energy are convertible, the equation relating them being the famous $E = mc^2$. In this equation E is the amount of energy, in joules, liberated by the annihilation of m

Proton-proton cycle (PPI)
(The dominant energy producing cycle in the Sun).

$$H + H \longrightarrow D + e^+ + \nu$$
$$D + H \longrightarrow {}^3He + \gamma$$
$${}^3He + {}^3He \longrightarrow {}^4He + 2H + \gamma$$

Secondary proton-proton cycle (PPII).

$${}^3He + {}^4He \longrightarrow {}^7Be + \gamma$$
$${}^7Be + e^- \longrightarrow {}^7Li + \nu$$
$${}^7Li + H \longrightarrow {}^8Be \longrightarrow 2\,{}^4He$$

Tertiary proton-proton cycle (PPIII).

$${}^3He + {}^4He \longrightarrow {}^7Be + \gamma$$
$${}^7Be + H \longrightarrow {}^8B + \gamma$$
$${}^8B \longrightarrow {}^8Be + \nu \longrightarrow 2\,{}^4He$$

CNO cycle.
(The dominant energy producing mechanism in massive stars).

$${}^{12}C + H \longrightarrow {}^{13}N + \gamma$$
$${}^{13}N \longrightarrow {}^{13}C + e^+ + \nu$$
$${}^{13}C + H \longrightarrow {}^{14}N + \gamma$$
$${}^{14}N + H \longrightarrow {}^{15}O + \gamma$$
$${}^{15}O \longrightarrow {}^{15}N + e^+ + \nu$$
$${}^{15}N + H \longrightarrow {}^{12}C + {}^4He$$

H	Hydrogen nucleus
D	Deuterium nucleus
Li	Lithium nucleus
He	Helium nucleus
Be	Beryllium nucleus
B	Boron nucleus
C	Carbon nucleus
N	Nitrogen nucleus
O	Oxygen nucleus
e^-	electron
e^+	positron
ν	neutrino
γ	gamma-ray photon

Figure 14.10 The energy-producing reactions in the Sun.

kilograms of matter. The speed of light, c, is 3×10^8 m/s. The energy–mass exchange rate is quite handsome. For every kilogram of mass converted, 9×10^{16} joules of energy are released.

Owing to the processes of fusion going on in its interior, the Sun is losing mass at the rate of 4 million tonnes every second to produce its energy output. However, there is plenty left and we think that the Sun will continue to shine for several thousand million years to come.

It should be emphasised that the details of the conditions and types of reaction occurring in the Sun are known only through theoretical models. Also, we have some indications that our models are insufficiently accurate. One type of particle that the Sun emits is the *neutrino*. A neutrino is massless and interacts only very weakly with matter. Most of the solar neutrinos pass straight through the Earth unimpeded. However, a small number of them can be made to interact with chlorine atoms. The result of an interaction is the conversion of a chlorine atom to one of radioactive argon.

In a classic experiment, which is still running today, Raymond Davis Jr had a 400 000 litre tank of the cleaning agent perchloroethylene installed 1.5 km below ground in a gold mine in South Dakota, in America. Special equipment purges the tank at regular intervals and measures the number of argon atoms created. The result is disturbing to the theorists. The Sun appears to be emitting neutrinos at only about a third of its expected rate.

A lower core temperature might explain the low neutrino flux, but this conflicts with other well-observed parameters. Perhaps the Sun's core undergoes fluctuations of activity. Certainly, the sunspot activity virtually ceased between the years 1645 and 1715, with spots rarely seen during this period. Attempts have been made to correlate changes in the climate on our planet to solar activity, though not with any great success. There are other possibilities. For instance, the neutrino might not be absolutely massless. Another idea is that the particular type of neutrino that is produced by the solar nuclear reactions might change in identity to one of the other two different types of neutrino that are known to exist. The correct explanation for the neutrino flux discrepancy is uncertain. The Sun is the nearest star to the Earth, but much remains to be learnt about it.

Questions

1 Describe in detail the structure and composition of the Sun.

2 Explain how a solar eclipse occurs. Show, by using a diagram, where on the Earth an observer would have to be to see (a) a partial eclipse and (b) a total eclipse.

3 Explain how you can **safely** observe the Sun, using a telescope.

4 Describe the appearance and nature of sunspots.

5   What is the *sunspot cycle* (also known as the *solar cycle*)? Describe how the cycle progresses, and explain the significance of a *butterfly diagram*.

6   Write an essay about the magnetic field of the Sun, emphasising the types of phenomena associated with, and influenced by, it.

7   What are *solar prominences*? Under what circumstances can they be studied without the use of special equipment? How are *filaments* related to prominences?

8   Write a short account of the solar corona.

9   Write a short essay about the *solar wind*. Include brief details about *solar flares*.

10   Describe the nature and structure of the *solar chromosphere*. Give details of any forms of observed phenomena that might be associated with the chromosphere.

11   Describe our current idea of how the Sun produces its power. What are *neutrinos* and what do they tell us about the processes occurring in the Sun?

Chapter 15

Electromagnetic Radiation

APART from meteorites and a few hundred kilograms of lunar rock samples, all our information about the physical Universe comes to us via electromagnetic radiation. Visible light is just one form of this radiation.

15.1 The Electromagnetic Spectrum

We can think of electromagnetic radiation as energy being conveyed in the form of a wave motion. Considering the analogy between these electromagnetic waves and, say, the waves on the surface of the sea, we can assert that the wavelength is the distance between two adjacent peaks. This is also equal to the distance between two troughs and the distance between the midpoint of one wave and the midpoint of the next complete wave (Fig. 15.1(a), *overleaf*).

Electromagnetic radiation can also be thought of as a stream of energetic, though massless, particles. We call these particles *photons*. In fact, neither description of electromagnetic radiation is entirely accurate. In certain circumstances it is more convenient to describe the radiation in terms of a wave and in others as a stream of particles! The truth lies somewhere between.

The different colours of visible light each have their own wavelengths and other forms of electromagnetic radiation, such as radio waves, X-rays, etc., are categorised by their own wavelengths. The complete range of wavelengths form the *electromagnetic spectrum* and this is illustrated in Fig. 15.1(b) (*overleaf*). Some readers might be surprised that all the radiations shown in the diagram – for instance γ-rays, visible light, microwaves – are really all the same thing. They only differ in their wavelengths and, because of this, in the ways they are created and detected. They all travel through a vacuum in straight lines with a speed of 3×10^8 m/s.

As the wavelength of the radiation gets less, so its photons become more energetic and penetrating. The relation between photon energy and wavelength is given by Planck's law. Expressed in the form of an equation this is:

$$E = hc/\lambda$$

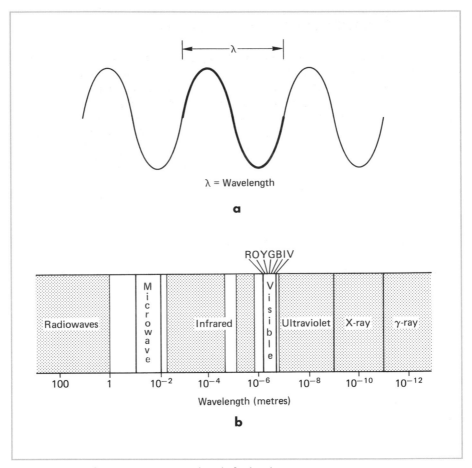

Figure 15.1 a Wavelength. **b** The electromagnetic spectrum.

where c = the speed of light,
   λ = the wavelength of the radiation in metres,
   E = the energy of the photon in joules.

h is known as Planck's constant. Its value is equal to 6.63×10^{-34} J s. Note that the unit of h is the joule second, **not** the joule *per* second.

The Earth's atmospheric mantle effectively absorbs most of the electromagnetic radiations coming from space, except for a few "windows" such as that centred on the visible part of the spectrum. The parts of the spectrum absorbed by our atmosphere are indicated, in a simplified form, on the diagram by the shaded areas. Certain narrow windows that exist have been left out of the diagram for clarity. In recent years astronomers have been anxious to study the celestial bodies in wavelengths other than just the visual. By setting up observatories on high mountain sites, such as that atop Mauna Kea in Hawaii, a few extra windows are opened up in the infrared and microwave regions. By using high-altitude balloons and aircraft astronomers have gained a little access to some extra ultraviolet wavelengths. Better results can be obtained only from rockets and satellites in space.

15.2 Black-Body Radiation

Most people are aware that bodies give off light when heated to high temperatures. The brilliance of the light and its colour largely depend upon the nature of the body. However, it is also true that all bodies approximate, at least to some degree, to something known in physics as a *black body*. A black body has certain definable characteristics. A perfect black body is a perfect emitter of radiation. In other words, any energy stored by a black body is freely radiated away in accordance with very definite physical laws. A black body is also a perfect absorber. This means that it is freely able to absorb incident radiation with no reflection at all.

A black body will emit radiation if it is at all above the absolute zero of temperature (the coldest it is possible to get according to physical theory). Also, all wavelengths of radiation are emitted from a black body at any value of temperature above absolute zero, but the amounts of energy radiated at any particular wavelength do vary with temperature. Figure 15.2 illustrates the distribution of radiation energy with wavelength from a black body at three different temperatures.

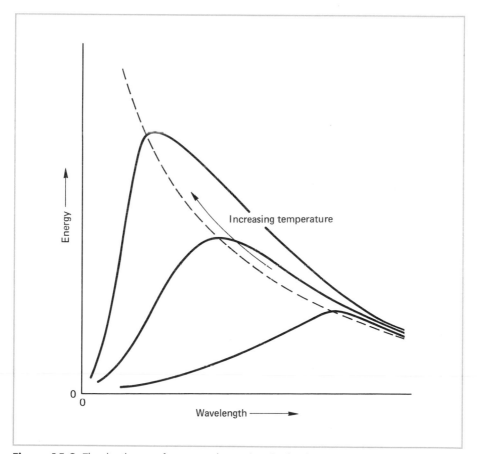

Figure 15.2 The distribution of energy with wavelength of radiation emitted from a black body.

It can be seen from the diagram that the wavelength of maximum energy emission shifts with temperature. The peak moves to shorter wavelengths as the temperature increases. In other words, a cold black body emits no visible light. A warm black body emits mainly in the infrared portion of the spectrum. Any hotter body will emit enough to be visible in the red part of the spectrum. As the temperature is further increased, so the body appears to brighten and become orange, then yellow. By this stage the temperature is some 5000 K. Increasing the temperature still further will make the black body white-hot, then greenish-white, then bluish-white and finally blue as the peak wavelength of emission shifts into the ultraviolet. The temperature corresponding to this is about 40 000 K.

The physical conditions on the photospheres of the Sun and the other stars are such that we can regard them as black bodies to a high degree of accuracy. So, from a star's colour we can determine its temperature. Note, however, that the **peak** wavelength of emission is just that and not the **only** wavelength at which the radiation is emitted. Thus the star colours are never vivid, or *saturated*. Instead they are usually just pastel shades. The Sun's photospheric temperature is 5800 K and so the wavelength of peak emission is about 5×10^{-7} m, which is why it appears yellowish-white. This is also the wavelength to which our eyes are most sensitive. This is no coincidence but is a natural consequence of adaptive evolution.

The relationship between the sum total amount of energy emitted by a black body over all wavelengths and its temperature is given by the Stefan–Boltzmann law:

$$E = \sigma T^4$$

where E is the total energy emitted per square metre of the surface of the black body per second and T is its absolute temperature. σ is known as Stefan's constant and its value is 5.7×10^{-8} W m^{-2} K^{-4}. So, once we have found the surface temperature of a star we can calculate the amount of energy being radiated by each square metre of its surface. This sort of information is very useful to astrophysicists.

15.3 The Beginnings of Spectroscopy

Spectroscopy started as a result of investigations by Isaac Newton in 1666. He used a glass prism to split sunlight into its component colours. He found that he could recombine the colours back into the normal colour of sunlight. He tried isolating the separate colours with a slit and then used another prism to see if the colours could be further split. They could not.

We now know that the prism splits up, or *disperses*, light into its component wavengths. The glass prism works because light is *refracted* (has its direction changed) by different amounts that depend upon the wavelength. Normal white light is a mixture of wavelengths (and thus colours). As white light passes through a prism the shorter wavelengths are refracted through a greater angle than the longer wavelengths. Hence the light is sorted into its colours, as illustrated in Fig. 15.3(a).

This forms the basis of the instrument known as a *spectroscope*, shown in Fig. 15.3(b). The light from the source to be studied passes through a narrow slit and the resultant divergent beam is rendered parallel by a collimating lens. The beam

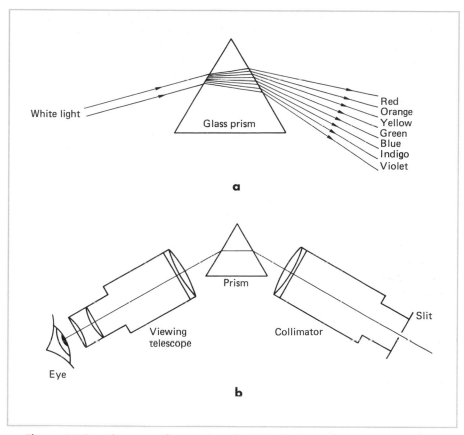

Figure 15.3 a The action of a prism on white light. **b** The simple prism spectroscope.

then passes through a prism (or a series of prisms) and the resulting dispersed beam is examined through a small telescope. This focuses the beam and presents it to the eye. In this way a series of images of the slit are formed, each for a different wavelength. The spectroscope has been developed to be one of the most powerful analytical tools available to the modern astronomer. Before we consider this development let us take a brief look at the types of spectra visible in such an instrument and their physical causes.

15.4 The Production of Spectra

The year 1859 saw the publication of the results of experiments carried out by Robert Bunsen and Gustav Kirchhoff. They worked together, sprinkling chemicals into the flame of Bunsen's newly invented burner and observing the colours produced through a spectroscope. They found that several types of spectra could be formed and their results laid the foundations that were necessary for our modern understanding.

A smooth rainbow of colours is known as a *continuous spectrum*. It is produced by any hot body that is made up of matter in relatively high density, such as in a liquid or a solid. A smooth black-body distribution is a continuous spectrum.

The situation changes when we consider the light produced by matter at low density – such as gases at low pressure. These produce *line spectra*. In order to understand how they do this we must first consider the structure of an atom. An atom of an element, say hydrogen, consists of a tiny positively charged nucleus surrounded by an electron (other elements have a greater positive charge on the nucleus and more electrons orbiting it). The electron can be thought of as a satellite moving round a planet. In practice the situation is not quite as simple as this. A better description of the electron is a cloud of charge that surrounds the nucleus in a standing-wave pattern.

A violinist can produce a standing-wave pattern by bowing one of the violin's strings. It appears as a stationary series of loops along the length of the string. If the violinist produces a pattern of just one loop along the length of the string, then it would be said that the string is vibrating in its "fundamental" mode. He or she could well get the string to vibrate in its "second harmonic". This is where two complete loops form along the length of the violin string (Fig. 15.4). More loops could be formed but, since the ends of the strings are fixed, it is impossible for the violinist to form two and a half, or three and a half, loops. Only a whole number of standing-wave loops are possible.

In the case of the electron the shape is more complex and is in three dimensions. However, there is a good analogy between the electron and the violin string. In neither case can anything other than whole numbers of standing-wave "loops" be accommodated. The "fundamental" standing wave pattern of the electron corresponds to its lowest possible energy level. The next possible standing-wave pattern (corresponding to two loops on the violin string) has a somewhat higher energy associated with it.

We have not the space for a full description of these electron standing-wave patterns, but here it will suffice to say that the electron can exist only in one of a number of discrete energy levels when it is trapped within an atom. It can exist in any of these levels, depending upon how much energy it has. It can accept energy to go from one energy level to a higher level. It can even give up energy to fall to a lower energy level, but it can never rest at any intermediate level.

With all this in mind we can explain the line spectra produced by a gas. If we supply energy to the gas by heating it to a very high temperature, or maybe drive an electric current through it by applying a high voltage, then we can cause some of the electrons surrounding the atomic nuclei to be excited to higher energy levels. In becoming excited the electrons take the energy being supplied. However, a sort of "what goes up must come down" principle operates and soon afterwards the excited electrons give up their energy and fall back to lower energy levels.

The crux of the matter is what happens to this discarded energy. **It is radiated away as photons of electromagnetic radiation**. The energy of a given photon is equal to the energy liberated by the de-exciting electron. This is equal to the difference in the values of the energy levels through which the electron falls. The wavelength of this photon depends on its energy and is given by Planck's law.

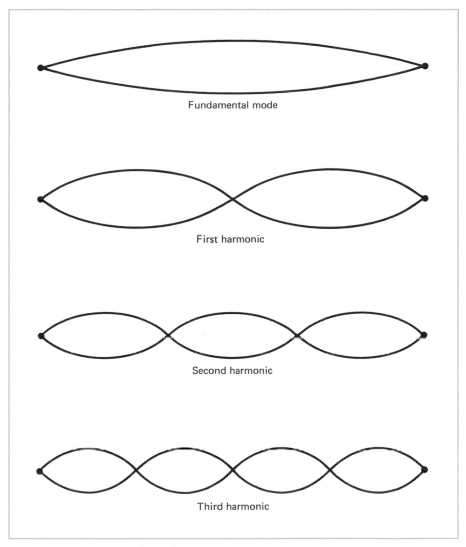

Figure 15.4 Standing wave patterns on a violin string.

In any given sample of the gas all the possible electron energy transitions will occur in a given small increment of time. In this way a characteristic series of wavelengths of radiation will be produced. If the light from the excited gas is viewed through a spectroscope then a continuous spectrum will not be seen. Instead, images of the slit will be seen only in certain wavelengths. A series of coloured lines will be produced, each in the position corresponding to its colour.

A spectrum of hydrogen consists of four visible lines: one in the red, one in the blue, one in the blue–violet and one in the violet. Various other lines occur in parts of the spectrum to which our eyes are not sensitive. Also, some electron energy transitions are preferred over others. This leads to spectral lines having differing intensities. The red line of hydrogen, known as Hα, is much more intense than the others for this reason.

Other elements have different arrangements of electrons in their atoms, and so each element produces its own characteristic spectrum. In this way an element can be identified in much the same way that a person can be identified from his fingerprint. If a certain person's fingerprint shows up in a room, then that person must have been in that room. If the lines of a certain chemical element show up in the spectrum of a source, then that element is present in that source.

15.5 Types of Spectra

When a spectrum is seen as a series of bright lines against a dark background it is known as an *emission spectrum*. It is also possible to get the appearance of a series of dark lines on a bright continuous spectrum. This is called an *absorption spectrum*. Most stellar spectra are of this type, so it is as well to consider how the situation arises.

Just as light may be given off when an electron becomes de-excited, so the incidence of light of exactly the right wavelength on an atom may cause an electron to jump to the corresponding higher energy level. If a little salt is sprinkled into a flame it is coloured golden yellow by sodium ions being released into the flame and becoming excited. If this flame is viewed in a spectroscope then the characteristic spectrum of sodium is seen, consisting of two bright lines very close together in the yellow part of the spectrum.

If the flame is exchanged for a powerful filament lamp then a continous spectrum is seen, produced by the hot metal filament. If the salt-sprinkled flame is then placed between the lamp and the spectroscope slit, so that the lamp is shining through the flame, something different is seen. The continuous spectrum is seen to be crossed by two dark lines in the yellow region – exactly where the bright emission lines were seen when the flame alone was viewed!

What is happening is that some of the light from the lamp, at just the right wavelengths, causes more energy jumps in the electrons surrounding the sodium nuclei. As a consequence this light is absorbed. When the electrons subsequently de-excite, the light that was sent by the lamp in the direction of the spectroscope is now reradiated in all directions. Hence the light at those specific wavelengths is depleted when it gets to the spectroscope. In this way dark absorbtion lines are seen on the bright continuum.

The single atoms that we have been considering so far form *line spectra*. Two or more atoms bonded together, in other words molecules, are capable of more complicated transitions and they tend to produce *band spectra*. Band spectra have the appearance of a series of bands, each sharp at one end but fading at the other. Closer examination reveals that the bands are in fact made up of numerous fine lines very close to one another. Like line spectra, band spectra can be seen in either emission or absorption. However, most molecules are broken down into their component atoms before any really high temperature is reached.

15.6 The Development of Astronomical Spectroscopy

Astronomical spectroscopy began in 1802 when William Wollaston noticed seven dark lines crossing the continuous spectrum of the Sun. He thought that they merely marked the boundaries between colours and he investigated no further. However, in 1814 Joseph von Fraunhofer also observed these spectral lines. He constructed improved equipment and found a lot more. Eventually he accurately mapped the positions of over 500 of them.

We now know that the dark lines are absorbtion lines produced by the solar chromosphere. The photosphere of the Sun is sufficiently dense to produce a continuous spectrum. However, the chromosphere is much less dense and its lowest layers are cooler than the photosphere. So we see the brilliant light from the photosphere through the less bright chromospheric layer. Hence the absorption lines. Figure 15.5 shows part of a spectrum of sunlight reflected from the surface of the Moon.

The British amateur astronomer William Huggins, who had an observatory in London, by 1860 was tiring of making conventional visual observations. He then heard of the investigations of Bunsen and Kirchhoff. Huggins seized on spectroscopy as a new technique to apply to observational astronomy. He mounted single-prism and multiprism spectroscopes at the focus of his 8 inch (203 mm) refractor. Like astronomers today, he even used a comparison spectrum. He did this by using an induction coil to produce high-voltage sparks between metal electrodes. If he used electrodes made of copper, then copper lines would appear in the spectrum. Iron electrodes caused the production of iron lines, etc.

In the late 1860s Huggins successfully used photographic plates to record spectra – the spectroscope had become the *spectrograph*. He made some very important discoveries about the natures of the celestial bodies, and it is fitting that he should be regarded as the father of astronomical spectroscopy.

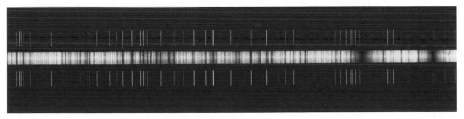

Figure 15.5 The spectrum of sunlight reflected from the surface of the Moon. Spectrum taken by the author using the 30 inch (0.76 m) coudé reflector, formerly of the Royal Greenwich Observatory, Herstmonceux. The above print is of a small section of the plate and covers a wavelength range of 20 nm in the near-ultraviolet part of the spectrum. The solar spectrum is the central stripe. It can be seen to be crossed by many dark absorption lines. The two broad lines at the right-hand end are due to the element calcium. The vertical lines above and below the solar spectrum are emission lines from a copper–argon comparison source, produced in the spectrograph at the time of the exposure.

15.7 Modern Spectroscopy

Prisms have been largely superseded by *diffraction gratings*, as the dispersing elements in spectrographs. A diffraction grating usually takes the form of a glass plate with many fine parallel rulings on its surface, often hundreds per millimetre. The ruled surface is then given a highly reflective coating, usually a film of aluminium.

The grating has an interesting effect on light falling on it. By a process of diffraction and interference (two manifestations of light behaving like a wave motion) the light is sorted into its component wavelengths. The different wavelengths are reflected off the grating at different angles. In other words, the grating disperses the light in the same manner as a glass prism, though by a totally different mechanism. The factor that makes diffraction gratings so useful is that they can produce a much higher dispersion (spreading of the wavelengths) and spectral resolution than is possible using prisms.

Modern high-dispersion spectrographs enable much quantitative analysis to be performed on the light sent to Earth from the celestial bodies. The pattern of spectral lines reveals the elements present in the body. Measurements of the intensities of the various lines also shows how much of the various elements are present. Measurements of the continuum allow the temperature of the source to be determined. Even the conditions of pressure and magnetic field strength present in the source can be ascertained from detailed measurements of the *line profiles* for very high-dispersion spectra.

The line profile is the way the intensity of a given spectral line varies with wavelength. At low dispersions the lines appear as just that – thin, dark (for absorption spectra) lines. With sufficiently high dispersion some structure can be seen in many spectral lines. The reasons for this are beyond the scope of this book, but suffice it to say that increased pressure and the presence of a magnetic field both tend to cause spectral lines to broaden. The first case is called *pressure broadening* and the second is known as the *Zeeman effect*. The two situations are distinguishable by their effects on line profiles. When reading the last chapter you might have wondered just how the magnetic field strengths over the surface of the Sun can be measured. This is one instance where the Zeeman effect has been put to good use.

Yet another application of spectroscopy is the measurement of the velocity of a celestial object towards or away from us along our line of sight. This is called *radial velocity* measurement and utilises the *Doppler effect*.

The Doppler effect is so familiar to us in everyday life that we seldom appreciate its significance. An example often quoted is of a train moving fast along a track and sounding its whistle, first approaching, passing, and then receding from a bystander on the side of the track. The bystander hears a significant drop in the pitch of the whistle as the train passes him.

Figure 15.6 illustrates the cause. As the train approaches, so the sound waves from the whistle are "bunched up" in front of it. The bystander experiences these sound waves of shorter wavelength as an increase in the frequency (pitch) of the sound. When the train is receding the waves are "stretched out", so seeming to the bystander to be of a lower frequency. The same effect occurs for electromag-

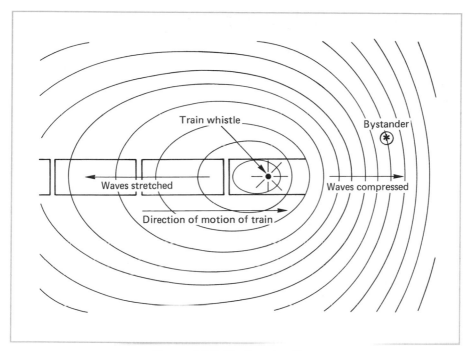

Figure 15.6 The Doppler effect.

netic waves. In this case the apparent change in wavelength can be measured directly as a shift in the positions of the lines in the spectrum of the moving source. If the source is receding then the apparent wavelengths of the lines are increased. This is called a *redshift*, since red light is at the long-wave end of the visible part of the spectrum. An approaching source then produces an apparent *blueshift* of the spectral lines.

In practice astronomers measure the wavelengths of the spectral lines from the astronomical body using a comparison spectrum (usually provided by a low-pressure gas discharge lamp in modern spectrographs) and these are compared with the laboratory-determined wavelengths. The radial velocity of the body can then be found from the following simple formula:

$$\frac{\triangle\lambda}{\lambda} = \frac{v}{c}$$

where λ = the laboratory-determined wavelength of a given spectral line,
$\triangle\lambda$ = its apparent wavelength shift,
c = the speed of light.
v is then the radial velocity of the body.

As illustrated in Chapter 4, modern technology has been readily utilised in spectroscopy. Electronic detectors are currently superseding the photographic plate and new types of instrument have been developed that allow high-quality spectra to be obtained in regions of the spectrum other than the visual.

Questions

1   Briefly explain the nature of *electromagnetic radiation*, and in the process define the term *wavelength*. Draw a diagram illustrating the full electromagnetic spectrum, giving the wavelengths of the various radiations. Include the colours of the visible spectrum, taking care to put them into their correct order.

2   Explain the effect which a glass prism has on an incident beam of white light. Show how such a prism is incorporated into a spectroscope.

3   Explain how *continuous spectra* and *line spectra* are produced. Explain how line spectra can be seen in either *emission* or *absorption*.

4   Give a brief account of the historical development of astronomical spectroscopy.

5   Explain how spectroscopy can be used to provide information on a whole range of physical conditions pertaining to celestial bodies.

6   The spectra of three stars X, Y and Z are photographically recorded with comparison spectra. These enable measurements of the apparent wavelengths of the spectral line Hα to be made for each star. The results are given in (a), (b) and (c) below. In each case work out the value of the radial velocity of the star and state whether the star is approaching or receding from the Earth. (The "rest" wavelength of the Hα line, as determined in the laboratory, is 6.563×10^{-7} m. The speed of light = 3×10^{8} m/s.)

(a)   Apparent wavelength of the Hα line from the spectrum of star X = 6.644×10^{-7} m.

(b)   Apparent wavelength of the Hα line from the spectrum of star Y = 6.522×10^{-7} m.

(c)   Apparent wavelength of the Hα line from the spectrum of star Z = 6.700×10^{-7} m.

7   Find out about, and write a short account of, how rainbows are produced. Include any diagrams that are necessary to aid your explanations. You should then be in a position to answer the following questions:

(a)   During some drizzly rain one day in midwinter the Sun appears through a gap in the clouds. It is noon. In what compass direction should you look if you expect to see a rainbow?

(b)   What are the uppermost and lowest colours in the primary bow?

(c)   What are the uppermost and lowest colours in the secondary bow?

Chapter 16

The Stars

GAZING up on a brilliantly starlit night, from a site well away from lights and buildings, fills me with a feeling of serenity. I can also sense the vastness of space and the immense power sources responsible for the shining stars. Subjective though these emotions are, I know that countless people have had similar experiences right back through the ages to the dawn of mankind. It is little wonder that astronomy is the oldest science.

16.1 The Natures and Brightnesses of the Stars

We have already considered the nature of one star in detail – the Sun. All the stars in the sky are, like the Sun, immense globes of incandescent gas. They appear much less bright than the our daytime star only because of their colossal distances from us.

Though the apparent magnitude scale has already been mentioned in passing, now is the appropriate time to give it a firm mathematical foundation. On this scale the brightness of a star is given by the following expression:

$$m_v = -2.5 \log I$$

where I is the apparent brightness of the star, in relative units, and m_v is its resulting apparent magnitude. The star Vega is the primary standard and is defined to have an apparent visual magnitude of 0.0.

The difference in apparent visual magnitude between one star and another, Δm_v, is then given by:

$$\Delta m_v = 2.5 \log(I/I')$$

where I and I' are the relative brightnesses of the stars, I being the intensity of the brighter star (which also has the **lower** magnitude number).

The brightest star in the sky (apart from the Sun!) is Sirius (see Fig. 16.1, *overleaf*), with an apparent visual magnitude of –1.5. On a really dark and clear night, well away from any sourcs of "light pollution", one can see stars down to

Figure 16.1 Sirius is the brightest star in this photograph, taken by the author.

magnitude 6.5. The sky then seems crowded with them. Dozens of bright stars, a larger number of fainter ones and an even greater number of still fainter ones can be seen. The dimmest stars appear to dust the sky. One might imagine that a million or more individual stars can be seen, but this is not so. Despite appearances, one can only see a maximum of about four thousand at any given time.

A telescope will reveal yet fainter stars. The bigger the aperture of the telescope, the fainter the stars that can be seen with it. An equation that gives a good guide to the limiting visual magnitude of a telescope is the following:

$$\lim m_v = 4.5 + 4.4 \log D$$

where D is the aperture of the telescope in millimetres. This equation is the author's own, but it is based on the results of an extensive practical survey carried out the by Bradley E. Schaefer, of the NASA-Goddard Space Flight Centre. Most of the older formulae give rather more pessimistic values for the limiting stellar magnitude visible with a given telescope. However, this equation, like the older ones, is not exact. It cannot be, since every different telescope will have its own value of optical efficiency. Also, the atmospheric conditions at the time of observation and the eyesight of different observers will differ markedly. If the seeing is too poor or the magnification used is too low then one will not be able to see stars as faint as that predicted. However, it is a fair estimate. Hence the limiting visual magnitude for an observer using a 6 inch (152 mm) reflecting telescope is 14.1, while the equation predicts a magnitude limit of 16.2 if he or she uses an 18 inch (457 mm) reflector under good conditions.

16.2 The Distances of the Stars

In the late 1700s the great observational astronomer Sir William Herschel considered that the stars were virtually equal in brightness, so that the fainter stars were always more distant than the brighter ones. However, by the time that he was nearing the end of his life, in the early 1800s, he was forced to change his mind owing to the increasing weight of evidence to the contrary. We can measure the distances of the nearby stars by the method of *trigonometrical parallax*. This is where a star's position is measured with respect to several others. As the Earth moves around the Sun, so a nearby star will apparently shift its position with respect to the more distant stars. Figure 16.2 (*overleaf*) shows the principle.

The extremes of position occur six months apart, since this is the period over which the Earth travels halfway round its orbit. Half the total angular shift of the star is the parallax. So far no star has been found with a parallax greater than 1 arc second. A 1 arc second parallax would mean that the star was a distance of 206 265 AU from us. This is 30 million million kilometres. Astronomers prefer to call this distance 1 *parsec*. The number of parsecs is found by taking the reciprocal of the number of arc seconds of parallax. For instance, a star that has a parallax of 0.5 arc second is 60 million million km, or 2 parsecs, from Earth.

Another popular unit of distance measurement is the distance travelled by a pulse of light in one year – the *light year*. A parsec is equal to 3.26 light years. The closest stars to Earth that astronomers have so far found, the two components of α Centauri and its gravitationaly bound neighbour Proxima Centauri, lie at a distance of 1.3 parsecs (4.3 light years). The difficulty of measuring tiny angular movements puts a limit of about 100 parsecs on the distance for which we can use parallax. This only covers a fraction of the stars which we can see with the naked eye. Luckily, astronomers have devised a variety of methods of extending distance measurements further out. These methods are considered later.

16.3 Absolute Magnitude and Distance Modulus

If we know how far away a star is and we measure its apparent brightness, then we can find its real luminosity. This is often expressed as its *absolute magnitude*. The absolute magnitude of a star is the apparent magnitude it would have if it was set at a standard distance of 10 parsecs from Earth. The Sun's apparent magnitude is –26.7, but its absolute magnitude is 4.8, so it would appear rather insignificant if it was placed at the standard distance of 10 parsecs away Absolute magnitude is denoted by M to distinguish it from apparent magnitude, m.

The quantity $(m - M)$ is useful, as it fixes the distance of a given star. It is given a special name: the *distance modulus* of the star. The equation is:

$$(m - M) = (5 \log d) - 5$$

where d is the distance of the star in parsecs.

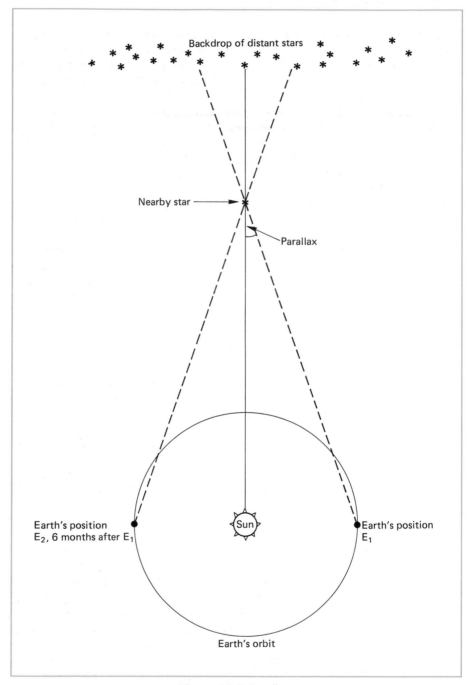

Figure 16.2 Parallax.

16.4 Star Colours

Look carefully at the stars and you will see that many of them show subtle, though definite, pastel shades of colour. On a summer evening one of the brighter stars in the sky is Vega. It shows a definite blue tint. Another bright star, Arcturus, is a pale orange–red. As discussed in Chapter 15, different star colours result from differing photospheric temperatures.

In order to classify star colours, astronomers measure their apparent magnitudes in different wavebands of the spectrum. They do this using combinations of photometric device and filter. The basis of the measures is the UBV system of Johnson and Morgan. In this system the ultraviolet (U), blue (B) and yellow (V for "visual") magnitudes are recorded. The system has been expanded to include red (R) as well as a variety of infrared wavebands. The difference in a star's magnitude in two wavebands is known as the *colour index* of the star. That normally quoted is the difference between the blue and yellow magnitudes: B–V.

Astronomers have defined the star Vega to have a blue magnitude of 0.0, the same as its visual magnitude. This means that it has the added distinction of having a colour index of 0.0. Redder stars than Vega have positive values of colour index while those that are more strongly blue have negative values.

16.5 The Constellations

To men in ancient times the stars seemed to form definite patterns in the sky. They arranged them into groups, the *constellations*. In most cases they were given names from ancient myths, such as Orion, the hunter who tangled romantically with the goddess Diana and was placed among the stars for his pains, and Andromeda, the maiden who was chained to a rock by the jealous sea god Neptune, to be devoured by a sea monster , but was rescued by the gallant Perseus. Ptolemy's great work *The Almagest* lists forty-eight constellations and these, with a few modifications and additions, form the basis of the constellations we know today. Astronomers use the Latin forms of constellation names, though their English equivalents together with the Latin names are listed in Table 16.1 (*overleaf*).

The constellation that is probably the easiest to recognise is Ursa Major (the Great Bear, though many people know it as "The Plough"), and it is shown in Figure 16.3 (*page 236*). Second most easily recognised is probably Orion (shown at the front of this book, in Fig. 1.1). However, when it comes to recognising the others many people have trouble relating what they see on a star-map to what is visible in the sky. Almost always this is because they imagine that the stars in a constellation form a rather small grouping and this is what they look for. In fact, most constellations are really quite big, sprawling, things. The photograph of Ursa Major, and Fig. 16.4 (*page 236*) which shows other constellations, have foreground buildings included in the view to give a better idea of how big constellations really look.

Table 16.1 The Constellations

Constellation	English name
Andromeda	Andromeda
Antlia	The Airpump
Apus	The Bird of Paradise (or the Bee)
Aquarius	The Water-bearer
Aquila	The Eagle
Ara	The Altar
Aries	The Ram
Auriga	The Charioteer
Boötes	The Herdsman
Caelum	The Sculptor's Tools
Camelopardalis	The Giraffe
Cancer	The Crab
Canes Venatici	The Hunting Dogs
Canis Major	The Great Dog
Canis Minor	The Little Dog
Capricornus	The Sea-goat
Carina	The Keel (of the ship Argo)
Cassiopeia	Cassiopeia
Centaurus	The Centaur
Cepheus	Cepheus
Cetus	The Whale
Chameleon	The Chameleon
Circinus	The Compass
Columba	The Dove
Coma Berenices	Berenice's Hair
Corona Austrinus	The Southern Crown
Corona Borealis	The Northern Crown
Corvus	The Crow
Crater	The Cup
Crux	The Southern Cross
Cygnus	The Swan
Delphinus	The Dolphin
Dorado	The Swordfish
Draco	The Dragon
Equuleus	The Foal
Eridanus	The River Eridanus
Fornax	The Furnace
Gemini	The Twins
Grus	The Crane
Hercules	Hercules
Horologium	The Clock
Hydra	The Sea-serpent
Hydrus	The Watersnake (or Small Sea-serpent)
Indus	The Indian
Lacerta	The Lizard
Leo	The Lion
Leo Minor	The Little Lion
Lepus	The Hare
Libra	The Scales
Lupus	The Wolf
Lynx	The Lynx

Table 16.1 continued

Constellation	English name
Lyra	The Lyre
Mensa	Table Mountain
Microscopium	The Microscope
Monoceros	The Unicorn
Musca Australis	The Southern Fly
Norma	The Rule
Octans	The Octant
Ophiuchus	The Serpent-bearer
Orion	Orion (the Hunter)
Pavo	The Peacock
Pegasus	The Winged Horse
Perseus	Perseus
Phoenix	The Phoenix
Pictor	The Painter
Pisces	The Fishes
Piscis Austrinus	The Southern Fish
Puppis	The Poop-deck (of the ship Argo)
Pyxis	The Mariner's Compass
Reticulum	The Net
Sagitta	The Arrow
Sagittarius	The Archer
Scorpius	The Scorpion
Sculptor	The Sculptor
Scutum	The Shield
Serpens Caput	The Serpent's Head
Serpens Cauda	The Serpent's Tail
Sextans	The Sextant
Taurus	The Bull
Telescopium	The Telescope
Triangulum	The Triangle
Triangulum Australe	The Southern Triangle
Tucana	The Toucan
Ursa Major	The Great Bear
Ursa Minor	The Little Bear
Vela	The Sails (of the ship Argo)
Virgo	The Virgin
Volans	The Flying Fish
Vulpecula	The Fox

Another problem arises for the tyro when the sky is too clear and transparent. If that sounds ridiculous take a look at Fig. 16.5 (*overleaf*), which shows the constellation of Cassiopeia in a field crowded with stars. A night of poor transparency shows only the brightest stars which form the distinctive "W" shape and the constellation is recognisable at a glance.

A common misconception is that the stars in a given constellation are physically related. A group of stars that have similar apparent magnitudes may well have vastly different absolute magnitudes and only look similar because of their very different distances from us. An apparently dim star may well be dim and

Figure 16.3 The constellation of Ursa Major. Photograph by the author.

Figure 16.4 The kite-shaped group of stars just to the upper-right of the house with its lights on is the constellation of Scorpius. The brightest stars of Libra can just be seen to its right and the long line of stars above Scorpius form part of Ophiuchus, and extends into Serpens Caput at the upper right. Photograph by the author.

Figure 16.5 How easily can you identify the rough "W" shape formed by the brightest stars of Cassiopeia among all the fainter ones in this time exposure photograph by the author? Nights of poor atmospheric transparency are often better for learning the main constellation patterns as they are then more easily recognised.

close to us. However, it might also be brilliant but very remote. Moreover, if we could shift our position in space by a few dozen parsecs then the patterns of the constellations would appear very different. To summarise, the constellation patterns are purely line-of-sight effects with no physical significance.

You might wonder why we retain the old constellation names. In fact they allow us to use a unified system of nomenclature for cataloguing the stars. On this system the brightest star of the constellation is denoted by the Greek letter α, the second brightest by β, the third by γ and so on through the Greek alphabet. Thus the brightest star in the constellation of Canis Minor is α Canis Minoris. The second brightest is β Canis Minoris, and so on. Unfortunately the system has not always been rigidly applied. Thus α Orionis is usually less bright than β Orionis (α is, in fact, variable in brightness).

Many of the brighter stars have been given popular names. For example α Orionis is Betelgeux, β Orionis is Rigel, α Lyrae is Vega, α Boötes is Arcturus, etc. Astronomers use other systems for identifying the stars that are fainter than those covered by this scheme, but these need not be discussed here.

16.6 Stellar Spectra

William Huggins was not the only pioneer of stellar spectroscopy, though he was the most important of the early observers. While Huggins observed the spectra of relatively few stars in great detail, Father Angelo Secchi examined the spectra of a great many stars in less detail. He achieved this by using a large, though thin, flint glass prism placed in front of the objective lens of a 9 inch (23 cm) refractor. With this he could see the spectra of several stars at once, though with a much poorer dispersion than Huggins obtained with his spectroscope.

Secchi eventually classified all stellar spectra into four types, though these are now only of historical interest. In 1890 E.C. Pickering devised a more precise scheme with an alphabetical sequence of eleven letters, A, B, C, D, etc., being used. Pickering's system survived for a while but some classes became unnecessary and the sequence was rearranged. Today the spectral sequence has become

W, O, B, A, F, G, K, M, R, N, S

(If you ever pass an astronomer's closed office door and you hear him saying "Wow! Oh Be A Fine Girl, Kiss Me Right Now, Sweetheart", you need not doubt that he has got his mind on his job. He will (probably) be remembering the order of the spectral sequence!)

Apart from type W, which is now regarded as a special case, there is a progressive change in the appearances of the spectra going along the sequence. In fact the basic types have been further divided into ten sub-groups. Thus type A0 is very similar to type B9 and A5 is halfway between types A0 and F0. Brief details are given in Table 16.2 and spectra of our Sun (spectral type G2) and Vega (A0) are shown for comparison in Figure 16.6.

We now know that the major differences in the spectra are due to temperature variations and not to differences in the compositions of stars. As an example, the chromospheric temperatures of N and S type stars are low enough to allow some very stable molecules, such as zirconium oxide, to form. So in the spectra of these stars we find the characteristic bands that are the signature of these molecules.

Table 16.2 Stellar spectra

Spectral type	Typical photospheric temperature (K)	Colour	Spectral characteristics
W	50 000	Blue	Wolf–Rayet; bright emission lines
O	50 000	Blue	Helium lines prominent
B	23 000	Blue–white	Helium and hydrogen
A	11 000	White	Hydrogen dominant
F	7 600	Yellow–white	Weaker hydrogen; calcium lines strong
G	6 000	Yellow	Metal lines prominent
K	4 000	Orange	Metal lines dominant; some molecular bands
M	3 500	Orange–red	Titanium oxide bands dominant
R	3 000	Red	Dominant carbon, carbon compounds and metal bands
N	2 600	Red	Ditto
S	2 600	Red	Zirconium oxide and titanium oxide bands dominant

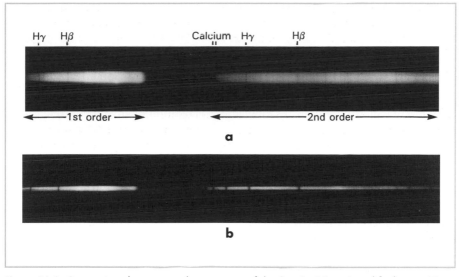

Figure 16.6 Comparison between **a** the spectrum of the Sun (a G2 star) and **b** the star Vega (an A0 star). Notice that Vega's spectrum is dominated by the Balmer series of lines of hydrogen, but many of the lines due to other elements visible in the solar spectrum are absent in Vega's spectrum. Spectra obtained by the author using the spectrograph on his $18\frac{1}{4}$ inch (464 mm) reflecting telescope.

These molecules cannot form in hotter stars, though zirconium and oxygen will be present in very similar proportions.

Also, the electron energy level transitions that generate specific lines are variously favoured at different temperatures. Referring again to Fig. 16.6, it is the higher photospheric/chromospheric temperature of the A0 star which causes it to display the "Balmer" lines of hydrogen so prominently, while inhibiting the formation of the lines due to other elements that are more evident in the solar spectrum. Yet the basic compositions of the Sun and Vega are very similar.

In fact the compositions of most stars are very similar to each other and to the Sun. So, the spectral sequence is a temperature sequence. Type O stars are hot and blue while type S stars are cool and red.

W stars are known as Wolf–Rayet stars, after the astronomers who were prominent in studying them. Wolf–Rayet stars are peculiar in that they show emission lines, rather than the dark absorption lines of other spectral types. They are discussed in more detail in Chapter 18. Some of the type N and S stars show enhancements in the strengths of their carbon lines. Astronomers think that these stars are immersed in clouds of carbon vapour and they are known as type C stars. In this way type C stars form a parallel branch of the N and S types.

16.7 The Sun – a Star Among Stars

Our Sun is a very ordinary star. It is one of about two hundred thousand million other stars that inhabit a vast system in space known as the Galaxy. To us the

Sun appears big and extremely bright but this is only because of our proximity to it. Rigel (β Orionis) has an absolute magnitude of –7.0. Comparing it with the Sun's absolute magnitude of + 4.8 means that Rigel is over 50 000 times as luminous as the Sun. Even this is not the upper limit. There are a few examples of stars that are the better part of a million times as bright as the Sun!

However, our Sun is not particularly puny. There are many examples of stars which are very much dimmer than the Sun. Some have been detected with only a millionth of the Sun's luminosity. How does a star's luminosity relates to its spectral type? The answer to this is a little complicated. A star of a particular spectral type might have any one of several values of luminosity. Thus astronomers have divided stars into *luminosity classes*.

Now, remember that the spectral sequence is a temperature sequence. Also, that Stefan's law relates the amount of energy emitted per unit area of a star's surface to the fourth power of its temperature. If astronomers determine a star's spectral type then they are also determining its photospheric temperature. Put another way, two stars of the same spectral type must have the same photospheric temperatures. Therefore if two stars of the same spectral type differ in brightness it must be because they have different surface areas. Indeed, the brighter star must have a larger surface area than the dimmer one.

If one star is 100 times brighter than another of the same spectral class, then it must have 100 times the surface (photospheric) area. Since the surface area of a sphere is proportional to the square of its radius, this means that the brighter star has 10 times the radius of the dimmer one. In this way the luminosity classes relate to the physical sizes of the stars.

Class I stars are *supergiants* with radii that are typically between 20 and 500 times as large as our Sun. Rigel, of spectral type B8, is a blue supergiant of about 20 solar radii. For historical reasons stars at the blue end of the spectral sequence are said to be of *early* spectral type and those at the red end are said to be *late*. In this luminosity class stellar radii increase, going from early to late spectral type. Betelgeux (α Orionis) is an M2 supergiant with a radius of nearly 500 times that of the Sun, though its lower photospheric temperature makes it only about a fifth of the brightness of Rigel. However, this is still 10 000 times as bright as the Sun!

Class II stars are *bright giants*. They mostly range between about 1000 and 10 000 times the luminosity of the Sun and have radii between 10 and 100 times that of the Sun. ϵ Canis Majoris is a blue example of this type of star. There are fewer examples of stars of this luminosity class than of the others.

Class III stars are the *giants*. They are mainly confined to spectral types G, K and M. In this type of star both the luminosity and the radius increase going from type G to M. Typical values are 30 to 1000 times the solar luminosity and 5 to 50 times the solar radius. Though most of the giants are yellow or red stars, blue examples do exist. γ Orionis (spectral type B2) is a case in point. Arcturus is more typical, being of spectral type K2. The blue giants are all much brighter than their red counterparts, approaching 4000 times the solar luminosity in the case of γ Orionis.

Class IV stars are known as *subgiants*. This is another thinly populated class, with typical luminosities ranging between 5 and 5000 times that of the Sun. Their luminosities increase from type K to type B, though their radii remain around 5 times that of the Sun for all spectral types.

Class V stars form the major class and are consequently known as *main sequence stars*. Most stars are of this class. Spectral type O main sequence stars have luminosities of about 300 000 times that of the Sun and radii about 20 times bigger. The radius and brightness of these stars decrease with later spectral types. A type M main sequence star has a brightness of the order of about 1% that of the Sun and a radius about one-third of the solar radius. The Sun is a typical main sequence star, of spectral type G2.

Class VI stars form a small group whose members are slightly smaller and less bright than the main sequence stars. They are known as *subdwarfs*. Most of them are of spectral types later than F.

Class VII stars form the final classification. These are the *dwarfs*, stars of very high density and tiny radius, typically one-hundredth that of the Sun. They have special significance for our ideas of stellar evolution, and there is more about them later in this book. They are all rather dim objects, usually having less than 1% of the Sun's luminosity.

Now that the luminosity classes have been covered in superficial detail, I can rectify an earlier oversimplification that I made in order to expedite the explanation of how astronomers deduce the sizes of stars from their intrinsic brightnesses and their spectral types. When I stated that the stars of a given spectral type are all at the same temperature I was being deliberately inaccurate in order to make the point clearly. In fact, the lower photospheric pressures in the larger stars actually cause them to mimic stars of a slightly different type.

As an example, an M0 supergiant star's photosphere is at a temperature of 3300 K. An M0 giant star has a photospheric temperature of 3600 K, while a main sequence star of the same spectral type has a photospheric temperature of 3900 K. Needless to say, astronomers make the appropriate corrections when calculating the temperatures and brightnesses of stars.

The luminosity class of a particular star is often given with its spectral type. As an example, the Sun is $G2_V$.

16.8 The Hertzsprung–Russell Diagram

In 1911 the Danish astronomer Ejnar Hertzsprung compared the spectral types and luminosities of the stars in several star clusters. In this way he could avoid any uncertainty in the distances of the individual stars, since the stars in a given cluster all lie at very nearly the same distance from us. He found a non-random distribution. The American Henry Norris Russell, working independently of Hertzsprung in 1913, found the same effect. A plot of intrinsic luminosity versus spectral type for a collection of stars is known today as a *Hertzsprung–Russell diagram*.

The H–R diagram, as it is usually called, has become a powerful aid to astrophysicists in many branches of stellar astronomy, particularly in the study of stellar evolution. Figure 16.7 (*overleaf*) is an example of an H–R diagram that might be obtained for a typical collection of stars in the neighbourhood of the Sun.

Note that if we can correctly identify the spectral type and luminosity class of a given star (there are small differences in the spectra that usually allow the

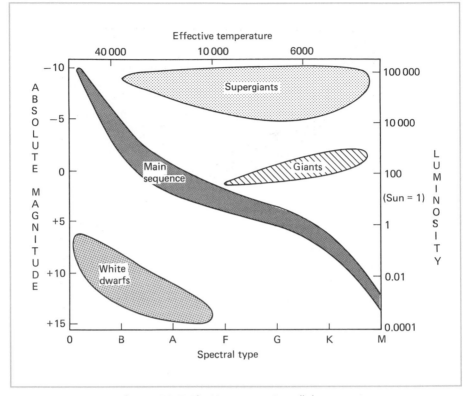

Figure 16.7 The Hertzsprung–Russell diagram.

luminosity class to be decided for a given spectral type), then we can assign a value of absolute magnitude to it. By measuring its apparent magnitude we can find the distance modulus, and hence the real distance, of the star. Of course, the whole diagram has to be calibrated by measurements of the absolute magnitudes of stars that are close enough to have their distances measured directly by the method of trigonometrical parallax, but this can be done. In this way we can determine the distances of stars that are far too remote to allow measurements of parallax.

One remarkable feature of an H–R diagram is that if stars of given radii are plotted on it, the result is a series of diagonal lines (see Fig. 16.8). This allows us to see at a glance how the radii of stars vary with spectral type and luminosity.

16.9 The Motions of the Stars

To a casual observer the stars appear rigidly fixed in their positions relative to each other in the sky. The constellation patterns do not appear to change from one year to the next. However, precise positional measurements do show that the stars undergo intrinsic changes in position. These are termed *proper motions* and are the result of the stars's real motions through space, relative to each other.

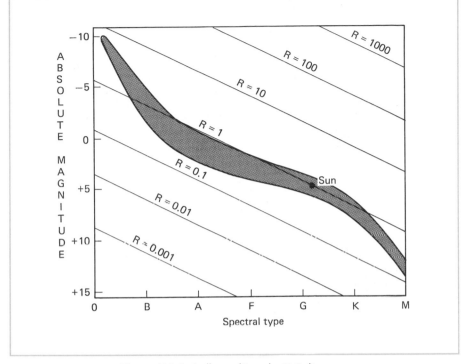

Figure 16.8 Stellar radii on the H–R diagram.

Most stellar proper motions are less than 1 arc second per year and are obviously smaller for the more distant stars. Barnard's Star, a 9.7 magnitude M5 dwarf, is an exception with an annual proper motion of just over 10 arcseconds. It is also, at 1.8 parsecs, the closest star to us, apart from the two components of α Centauri, Proxima Centauri and, of course, the Sun.

Measuring proper motions is not an easy affair. Despite the smallness of the positional changes, the astronomer has to disentangle the real proper motions from the other causes of the changes of a star's position. Nutation and precession (caused by the cyclic changes in the direction of the Earth's spin-axis) have to be taken into account as well as the effect of parallax, which is due to the Earth's orbit around the Sun. Instead of moving in straight lines, the stars showing proper motion move in wavy tracks due to parallax. Refraction in the Earth's atmosphere is yet another problem. This causes all the stars to be systematically increased in altitude by an amount that varies with the altitude (being greatest for a star near the horizon).

However, despite these complications, stellar proper motions can be and are being measured. If the distance to the star is known then its actual tangential velocity (across our line of sight) can be calculated. If these results are combined with radial velocity measures made spectroscopically then the two velocities can be added vectorially to give the star's real velocity in space, as well as its direction of travel. In the vicinity of the Sun most stars move with independent velocities ranging from a few kilometres per second to a few tens of kilometres per second.

16.10 Binary Stars and Stellar Masses

Look at the second-magnitude star called Mizar in the constellation of Ursa Major and you will see that it has a fifth-magnitude companion roughly a quarter of a degree away from it. This star, Alcor, is not associated with Mizar but merely lies in nearly the same line of sight. Even a small telescope reveals that Mizar is itself double. The two components are separated by about 14 seconds of arc. This is a case where the two components really are associated. They both revolve about the common centre of gravity of the system. Stars which appear close together in the sky are known as *double stars*. Some double stars are purely line-of-sight effects but many are systems mutually bound by gravity. These are known as *binary stars*.

Another example of a binary star is Albireo in the constellation of Cygnus. This is another pair that is easy to resolve, having a separation of 35 arc seconds (see Fig. 16.9). The brighter component (magnitude 3) is a golden yellow colour which contrasts beautifully with the fifth-magnitude blue companion.

Not all binary stars are as easy to resolve as Mizar or Albireo. Some require powerful telescopes and the very best viewing conditions. The components of a binary star might appear close because the system is remote, or it might be that the two components are actually close to each other. Sometimes the orientation of the stars causes them to appear close together from our viewpoint.

Interferometric techniques can be, and are being, used but spectroscopy also provides a powerful method for investigating very close binaries. The spectrum of a binary star is the composite of its individual components. As the stars revolve about their barycentre, so one star will be approaching us while the other is receding. Thus the light from one star will be blueshifted, while the other will be

Figure 16.9 The double star Albireo (β Cygni) photographed by the author, using the 36 inch (0.9 m) Cassegrain reflector, formerly of the Royal Greenwich Observatory at Herstmonceux.

redshifted. This results in an apparent "doubling up" of the spectral lines – one set from the receding star and the other from the approaching star. In many cases the spectral types of the components can be identified from the composite spectrum.

Stars too close for conventional visual observation but revealed by spectroscopy are known as *spectroscopic binary stars*. To find an example of these we have to look no further than Mizar. It turns out that each component is itself a spectroscopic binary. In other words, Mizar is a quadruple system. Binary stars that are observed for long enough show perceptible changes caused by their orbital motions (see Section 5.5). We can find the distances of many of these stars by the parallax method or from their spectra (and hence luminosities). In these cases we can use Kepler's laws and Newton's laws to derive the masses of the components. In this way astrophysicists can build up a picture of star masses and determine how they vary with different types of stars.

It turns out that stars vary relatively little in mass, the vast majority having masses between one tenth and one hundred times that of the Sun. A star's mass determines how it evolves and what sort of life it leads. These matters will be discussed in the following chapters.

Questions

1 Explain the terms *apparent magnitude* and *absolute magnitude*, related to a star. A star, X, has a magnitude of 5.6, while another, Y, has a magnitude of 2.4. What is the ratio of the brightnesses of the stars and which is the brighter?

2 (a) What is the magnitude of the dimmest star an observer can see if he is using a 200 mm aperture reflecting telescope? (b) What minimum size of telescope is required to see stars of magnitude 15.8?

3 Explain how the distances of some stars can be found by using the method of *trigonometrical parallax*. Define the *parsec* and the *light year*. Why is the method of parallax only useful for stars closer to us than about 100 parsecs?

4 The star Regulus has an apparent visual magnitude of 1.33. It is also known to have an absolute magnitude of –1.0. (a) What is the value of the distance modulus of this star? (b) How far is this star away from us? (c) What value of parallax would you expect this star to show? (d) What is the star's value of luminosity, compared with the Sun? (The Sun's absolute magnitude is 4.8.)

5 Write a short essay about star colours, explaining in the process the meaning of the term *colour index*.

6 Write a short essay about the constellations and stellar nomenclature.

7   Outline the modern scheme of the spectral classification of the stars. What is the main physical significance of the spectral sequence?

8   Explain how the size of a star can be found from a knowledge of the star's absolute magnitude as well as its spectral type.

9   Give an account of the *luminosity classes*, which astrophysicists have used to classify stars. Include data on the brightnesses and sizes of the stars in each classification.

10   Write a short essay on double stars, explaining how binaries can be used to provide information on stellar masses.

11   Explain how astronomers can deduce the speed and direction of travel of a star, relative to our Sun.

12   Sketch a Hertzsprung–Russell diagram, labelling the axes, and on it mark the position of the Sun and the star Vega (the appropriate data can be found in this chapter). Sketch the main-sequence line on the diagram. The star Betelgeux has a spectral type and absolute magnitude of M2 and −5.6. Plot Betelgeux on the H–R diagram and briefly explain its position on it.

Chapter 17

Gas, Dust, Stars and Variable Stars

IF an extraterrestrial visitor were to take a walk through the streets of a busy town on a Saturday morning he/she/it (let us say "he") would see a variety of people. Some would be very small, being seated in wheeled chairs and pushed along by larger specimens, or bodily carried by these larger specimens. There would be people a whole range of sizes and with a wide range of physical characteristics. The smallest people tend to have unlined faces and strongly coloured hair. The full-sized people tend to be less active (walking steadily rather than running and jumping) and many have greying hair and lines on their faces. Some people have very wrinkled faces and most of these have white or nearly white hair. They often move very slowly, perhaps walking with a stoop and with the aid of a stick.

If the extraterrestrial visitor made the assumption that all these people of different sizes were examples of the same species, then he might reasonably suppose that he was witnessing examples of people at differing stages in their life cycles. By studying a cross-section of individuals he would probably be able to describe the major features of a human being's development from infancy to old age.

Astronomers are in the same position as our hypothetical extraterrestrial visitor. They see a variety of objects in the sky and try to construct a sensible scenario for the way each type of object fits into the scheme of things. Consequently, they now have a fairly detailed idea of how a star evolves from its birth to its death and how a large variety of celestial objects fit into the story.

17.1 When Astronomers Got It Wrong

Sherlock Holmes famously remarked that "When all that is probable has been eliminated whatever remains, however improbable, must be the truth." In a way, this is relevant to the way astronomers have arrived at their present understanding of stellar evolution. Certainly this branch of astronomy would make a fitting subject for Conan Doyle's fictional detective.

Astrophysicists in the nineteenth century were becoming uncomfortably aware that they could not explain the source of the enormous amounts of energy radiated by the Sun and other stars. Lord Kelvin showed that a star could not be burning fuel in the same way as a coal fire, since it would have exhausted all its reserves in a few thousand years.

Near the end of that century astrophysicists began to realise that a large amount of energy is stored in any body because of its gravity. This led to a proposal by Sir Norman Lockyer in 1890. He thought that a star would begin life by condensing from a cloud of interstellar gas and dust. Under its own gravity it would form a spherical body, heating up as it contracted, as a consequence of the liberation of gravitational potential energy. Eventually it would be hot enough to glow visibly. The star would then be a large, bloated, body – a red giant. The star would then continue to contract slowly and steadily increase its temperature and brightness. As the temperature of the star increased so its colour would change: first red, then orange, yellow, white, greenish, bluish, then blue. By then the star would have completed most of its contraction and would then begin to cool off as a dwarf star, going through the colours in reverse. Eventually the star would become too cool to be visible.

Lockyer's scheme is wonderfully straightforward and easy to understand. However, calculations showed that a star like the Sun could only last a few tens of millions of years, while geologists maintained that the Earth was thousands of millions of years old. Lockyer's scheme fell.

The early years of the twentieth century saw the discovery of radioactivity and the beginnings of atomic physics. Electrons, protons and neutrons were discovered and recognised as particles that make up atoms. The year 1913 saw Henry Norris Russell put forward an idea that under the conditions inside a star protons would collide with electrons and mutually annihilate each other (they were known to be oppositely charged). In this way the matter inside the Sun and the stars would be gradually converted to energy.

Russell's idea certainly solved the longevity problem. In fact, it solved it too well. Under this scheme the stars could last for more than a thousand times the age of the Earth! Moreover, physicists soon understood atomic particles well enough to realise that electrons and protons would not annihilate each other if they met. The following years saw immense advances in atomic physics. Physicists began to understand that under certain conditions atomic nuclei could undergo changes that would release vast amounts of energy. Some of these changes could occur through the fragmentation of nuclei. This is *atomic fission*. Other changes could occur when nuclei were built up as a result of collisions with smaller particles. This is *atomic fusion*.

We now understand that the Sun and the stars shine mainly because of fusion reactions in their cores. Carl von Weizsäcker and Hans Bethe were the astrophysicists mainly responsible for pioneering our modern ideas. They worked independently but came up with similar conclusions in 1938. However, Lockyer was right about one thing. The stars are born from the condensation of interstellar matter and this is where we begin our survey of our modern ideas about stellar evolution.

17.2 The Interstellar Medium

In the region of our Sun, the stars are separated from each other with an average distance of around 2 parsecs. This space is not empty but is filled with extremely tenuous gas and dust. Moreover, this gas and dust is not evenly distributed but is very "lumpy".

Radio observations made at a wavelength of 21 cm reveal vast clouds of hydrogen gas with masses ranging from one tenth the mass of our Sun to over 1000 times the solar mass. The atoms in these clouds are typically at temperatures of around 80 K and the density of the gas is around 50 million particles per cubic metre. This may sound dense, but 50 million hydrogen atoms have a mass of only 8×10^{-17} kg.

There are also more massive clouds of molecules, mainly hydrogen, H_2, though with traces of more complex species. As revealed by millimetric infrared observations, these clouds are very cold. Temperatures below 10 K are typical. These molecular clouds tend to be very big, often having masses the better part of a million times as large as the Sun. Dozens of different molecular species have been identified, most of them compounds of hydrogen, carbon, oxygen and nitrogen.

Between them the atomic hydrogen clouds and the molecular clouds do not fill up all the space between the stars. A recent surprise is the discovery that the remaining space is filled up with a low-density (around 3000 particles per cubic metre) gas of very high temperature (about 1 million K). The discovery arose from ultraviolet and X-ray studies made by using artificial satellites. It now seems that this high-temperature gas permeates the entire Galaxy. Observations reveal that this gas is not steady, but rather seethes with turbulent motion. The free electrons in this turbulent plasma (neither molecules nor atoms can survive intact at these temperatures) cause the radio signals from distant sources to scintillate in the same way that the Earth's atmosphere causes the stars to appear to twinkle.

As well as the major gas component, about 1% to 2% of the interstellar mass is in the form of solid "dust" particles. As far as we can tell from their effects on visible light and their emissions in the infrared, the particles are mainly metallic silicates (aluminium, magnesium, etc.) and carbon (probably in the form of graphite). They also appear to have various gases frozen onto them. The particles are very small. They range in size from a thousandth of a micrometre to a few tenths of a micrometre. Astronomers estimate that about half of the light from the stars in our Galaxy is absorbed by these dust particles. In the process, the particles are heated. They reach an equilibrium at a temperature of about 30 K, when they then emit radiation in the far infrared part of the spectrum. At this temperature the particles are emitting energy at the same rate that they are absorbing it. This situation is called *thermal equilibrium*.

The interstellar dust does not absorb all wavelengths of radiation equally well. The short wavelengths are absorbed more than the longer wavelengths. This causes the light from distant stars to be both dimmed and apparently reddened.

So far we have not been able to sample the interstellar medium directly, even by using spacecraft. The solar wind sweeps back the interstellar medium to well beyond the orbits of the planets. Other stars must do the same thing – in effect blowing bubbles in interstellar space!

17.3 Gaseous Nebulae

On a clear night and away from bright lights, look about 5° south of the three bright stars that form the "belt" of Orion and you will see what appears to be a dim misty patch. Figure 1.1 shows it quite well. Through binoculars it can be seen much better as a great luminous cloud in space. Figure 17.1 is a close-up photograph of the same region, obtained using a telephoto lens, and shows an area of sky similar to that you might see in low-power binoculars. Figure 17.2 provides a splendid view of the nebula photographed by a large telescope. Because the brightness of the nebula varies greatly from its innermost to its outer regions, further photographs (Fig. 17.3(a) to (d), *overleaf*) are necessary to reveal structure in the innermost regions, which appear bleached out in Fig. 17.2.

The sky is full of such objects. The great French comet hunter Charles Messier published a list of 103 objects in the sky that he thought might be confused with a comet. Not all these objects are gaseous nebulae and some are very unlike comets in appearance. The Great Nebula in Orion, mentioned above, is on his list as entry number 42. We still use Messier's catalogue numbers today. The Great Nebula in Orion we know as Messier 42, or just M42.

The indefatigable William Herschel considered that Messier's list was inadequate, and he set himself the task of "reviewing the heavens" by sweeping the sky with his telescopes. In the process he discovered nearly two thousand more

Figure 17.1 Close-up of the region below the "belt" of Orion, showing the Great Nebula, M42. Photograph by the author.

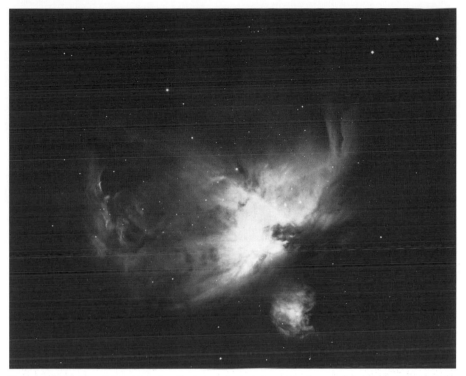

Figure 17.2 The Orion Nebula. Photographed at the prime focus of the Isaac Newton telescope, formerly at Herstmonceux, England. RGO photograph.

nebulae, though not all of these are the gaseous nebulae that we are at present considering. Today, Herschel's catalogue numbers have been superseded by the *New General Catalogue*. In this the Great Nebula in Orion is known by the number NGC 1976.

Visible nebulae are concentrations of the interstellar medium rendered so because of the way they affect, and are affected by, starlight. They are generally classed into one of three types: *emission*, *absorption* or *reflection* nebulae. Reflection nebulae are visible because they scatter (rather than simply reflect) the light of nearby stars. Our atmosphere scatters sunlight in the same sort of way. The shorter wavelengths of visible light are scattered more than the longer wavelengths. This is what makes the daytime sky blue, because the blue light from the Sun is preferentially scattered out of the arriving Sunlight. As a result we see the Sun as slightly redder than we would if we could see it above the Earth's atmosphere. This is most obvious when the Sun is seen through the greatest length of atmosphere, at sunset. This is the reason for interstellar reddening. It also results in reflection nebulae being blue, as is well shown in colour photographs.

Absorption nebulae are dark because they have no nearby stars to illuminate them. They betray their presence by absorbing the light from background stars. They give the visual impression of vast empty lanes, or holes, in space devoid of stars.

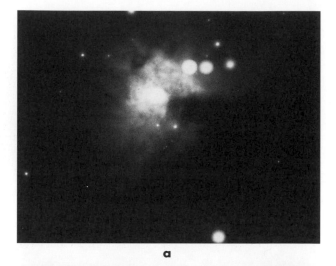

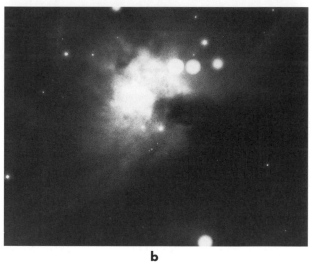

Figure 17.3 The Orion Nebula, M42, photographed by the author using the 26 inch (66 cm) astrographic refractor, formerly of the Royal Greenwich Observatory at their Herstmonceux site and pictured in Figure 3.15 earlier in this book. **a** to **d** show progressively fainter parts of the inner structure of the nebula.

Emission nebulae shine because their gas is excited into emission by the short-wave radiations emitted by very hot stars imbedded within them. Since the nebulae are composed mostly of hydrogen gas, the light they emit is mainly that characteristic of this element. For this reason they are also known as *HII regions*, since the hydrogen gas is ionised by the short-wave radiations from the hot stars. The dominant emission of hydrogen in the visible part of the spectrum is the Hα line, in the red part, with the blue Hβ line second in order of strength. Colour photographs of emission nebulae reveal their characteristic red/magenta colour.

Popular books on astronomy often mention that an observer using a telescope visually cannot see any of the colours of nebulae and photographs are needed to bring them out. This is completely true for the vast majority of nebulae. The reason is that the light levels are so low that the observer's eye is working in its

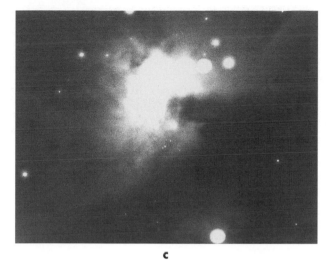

c

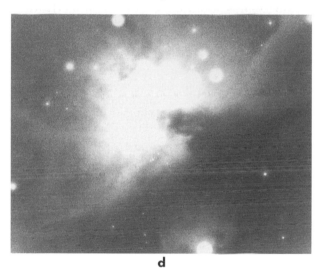

d

mode where it is very sensitive but can only record in monochrome. However, I find that I can see coloured tints in one or two of the brightest nebulae by using a large enough telescope.

My $18\frac{1}{4}$ inch (464 mm) reflector brings out slight bluish and rose-red tints in M42, on good nights. However, the colours are not strong and it may be that my eyes are more colour-sensitive than most. A few people with whom I have observed have seen the colours in M42 but others have been unable to see them. Some people say that the nebula looks green. Certainly the green spectral lines of oxygen are present in the light from the nebula, but the Hα and Hβ lines are most dominant. This is a sharp reminder that human perception is highly subjective.

M42 is a good example of all three types of nebula in one. The major part of the visible nebula is excited into emission by four stars, of spectral types O and B, known as the "trapezium" because of the appearance of the group in a telescope. Reflection nebulosity is visible near the stars in the brightest part of the nebula and dark absorption nebulosity is also visible in front of it.

Very long-exposure photographs of Orion show that M42 is just one bright concentration in a large and complex nebula that spans the whole constellation. Infrared images show the cloud well and reveal several "bright spots" not shown in visible wavelengths. These are possible sites of star formation.

17.4 Stellar Birthplaces

Dense molecular clouds are the regions where new stars form. The general scenario, as far as astronomers understand it, is the same as that described in Chapter 14 for the birth of the Sun. The large clouds fragment into many self-gravitating cloudlets, which then further contract to form stars at their central concentrations.

However, astronomers are becoming aware that the precise mechanisms which operate in star formation complicate the above scheme. There is a complex interplay of gravity, electromagnetic radiation and the temperature of the contracting cloud of gas and dust, as well as the magnetic fields generated within it. For this reason, more than one set of initial conditions may spark the formation of stars. Stars are formed sometimes in massive molecular cloud complexes and sometimes in smaller units. Many of the small, dark nebulae we call *Bok globules* are known to be contracting and will ultimately give rise to stellar birth.

One fact that should be apparent from this is that stars do not tend to form singly. A whole group of them tends to form at more or less the same time. These give rise to the formation of *open star clusters,* groups of maybe a few dozen to several hundred stars in the same small region of the sky. One magnificent example of an open star cluster is that in the constellation of Taurus, the Pleiades. To the naked eye the Pleiades appear as a little group of stars, covering an area about four times that of the full Moon. On an average night I find that I can count 6 Pleiads (the singular name given to the component stars), but I can see a few more on nights of really excellent transparency. A pair of binoculars reveals many more. In fact, there are over 400 individual stars composing this cluster (some recent studies indicate that there might be ten times this number), though many can be seen only by using a large telescope.

The cluster is 126 parsecs from us and is over 4 parsecs across. Figure 17.4 shows a photograph of the region obtained using a telephoto lens. Long exposure-photographs through a telescope, such as that shown in Fig. 17.5, show wreaths of reflection nebulosity surrounding the stars. This is thought to be material left over from the initial cloud that created the stars, roughly 20 million years ago. At least that is what the older textbooks state. A recent study indicates that

Figure 17.5 The Pleiades. A photograph taken with a 30 inch reflector in 1918. Note the nebulosity formed around the stars. This is material left over from the formation of the cluster. RGO photograph.

Figure 17.4 The region of the Pleiades star cluster, in Taurus, photographed by the author using a telephoto lens.

Figure 17.6 The Beehive star cluster, M44, in Cancer, photographed by the author using a telephoto lens.

the original nebulosity has probably been replaced as fresh material moves through the cluster.

Star clusters are not permanent constructions. Over a period of time the proper motions of the individual stars cause star clusters to disperse. The Beehive Cluster, M44, in Cancer is shown in Fig. 17.6. It is older and more sprawling than the Pleiades.

A much older cluster (about 4000 million years old) is the Hyades (Fig. 17.7), which happens to be in the same constellation as the Pleiades. The most prominent stars form a "V-shape on its side" with the bright, reddish star Aldebaran. Aldeberan is not actually a cluster member but lies in the same general direction. The stars of the Hyades are much more dispersed and individual cluster members can be found over a very wide area of the sky.

17.5 Stellar Birth and Infancy

An embryonic star can be said to exist as a separate identity when a core begins to develop in its contracting cloudlet. As material piles down onto this core so its temperature increases. When the core temperature reaches about 10 million K nuclear reactions begin. The star then strives to reach an equilibrium with the force of gravity (tending to make it contract) and gas pressure (tending to make it expand). At first the star is very unstable and erratically varies its luminosity. A well-studied example of this process at work is the star T-Tauri.

Figure 17.7 The Hyades star cluster in Taurus, photographed by the author using a telephoto lens.

Stars that change their brightnesses are known as *variable stars*. Around 20 000 examples of variable stars are known to us and there are many reasons for variability. Newborn stars that "hiccough" are known as T-Tauri variables, after their prototype. After passing through the T-Tauri stage stars settle down to a generally more sedate existance. On a Hertzsprung–Russell diagram this is represented by the *zero-age main sequence*. The mass of the star determines its luminosity and spectral type, and so its position on the main sequence. This is illustrated in Fig. 17.8 (*overleaf*) and arises because a star of larger mass has a stronger gravitational pull and so can only reach stable equilibrium if it is hotter.

Most stars spend the larger part of their active lives on the main sequence but just how long a star remains there depends upon its mass. These matters and what happens next in the life of a star are discussed in the next chapter.

17.6 Extrinsic Variable Stars

T-Tauri stars change their apparent magnitudes because of real changes in the luminosity of the stars. There are a great many of these *intrinsic variable stars*. There are some others which only mimic intrinsic variables because of geometric factors. *Eclipsing binary* stars are the main examples of these *extrinsic variables*.

In an eclipsing binary one star orbits another in a binary system whose orbital plane is at least approximately in our line of sight. With every orbit at least one of

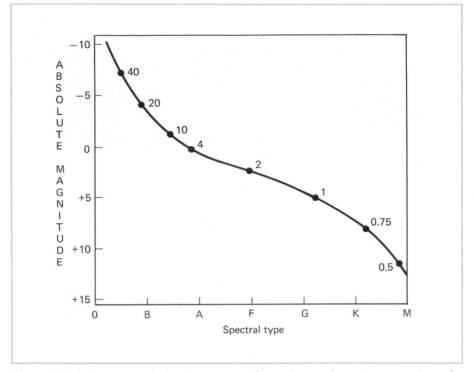

Figure 17.8 Star masses on the main sequence. The numbers on the main sequence line refer to the masses of the stars in Solar units (Sun = 1) in the positions shown.

the stars (usually both) partially or wholly hides the other for a short time. In this way the total amount of light sent to us from the system varies in a very regular manner.

The star known as Algol, in the constellation of Perseus, is the most famous example of this type of extrinsic variable, but there are many others. In this system the brighter star is a giant of spectral type B8 and absolute magnitude −0.2. This star has a mass five times that of our Sun and a diameter 3.2 times the solar diameter. The fainter star is also a giant, this time of spectral type F8 and absolute magnitude +2.5. It is estimated to have a mass equal to our Sun, though with a radius 3.7 times as large. The two components orbit at a distance of about 10 million km with a period of 2.87 days.

The combined apparent magnitude of the stars is 2.1 (from the Earth they cannot be seen as separate). The system remains at this brightness for most of the time but as the brighter component passes behind the fainter, so the apparent magnitude drops to a minimum of 3.3 over a period of five hours. During the next five hours the apparent magnitude increases to 2.1 again. It remains at this level until the next eclipse just over two and a half days later. Actually, there is a secondary minimum that occurs when the brighter component passes in front of the fainter star but the decrease in brightness is only 0.05 magnitude this time. The variation of apparent magnitude with time can be plotted as a graph known as a *light curve*. A light curve for Algol is shown in Fig. 17.9 with the secondary minimum greatly exaggerated for clarity.

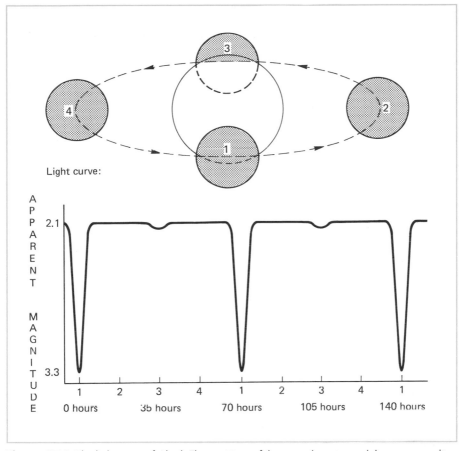

Light curve:

Figure 17.9 The light curve of Algol. The positions of the secondary star and the corresponding magnitude values are indicated by the numbers in the upper and lower parts of the diagram.

17.7 Intrinsic Variable Stars

Intrinsic variable stars can be divided into three types: *rotating variables*, *eruptive variables* and *pulsating variables*. Rotating variable stars have large areas on their photospheres which are either hotter or cooler than the average. As the star rotates, so its patchy surface causes apparent variations in brightness. Eruptive variables are characterised by erratic and unpredictable changes in luminosity. T-Tauri stars are just one class of eruptive variable. SS Cygni and U Geminorum stars undergo nova-like outbursts, generally every few months.

Pulsating stars vary their brightness because of cyclic changes in their size. Just as the air in an organ pipe can be made to vibrate at its resonant frequency, or multiples of this frequency, so can the material in a star provided that some mechanism is present to excite the oscillations. Theorists think that layers of material, particulary helium, can develop in a star under certain conditions. This layer is opaque and tends to act like a valve, causing a buildup of heat below the layer. The gas below the layer then expands, forcing the whole of the star to swell up. In the

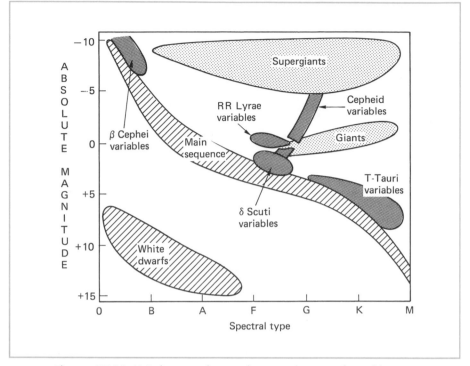

Figure 17.10 H–R diagram, showing the principle areas of variable stars.

process the opaque layer thins and allows the built up heat to escape. The star then contracts, the opaque layer thickens and the whole cycle starts all over again.

Variable stars do not have a random distribution on the Hertzsprung diagram, as Fig. 17.10 shows. This tells us that certain combinations of a star's physical parameters favour variability rather than stability. Variable stars are classified by their light curves. A few of the many types are shown in Fig. 17.11. and the main classes of pulsating variable are briefly described as follows.

RR Lyrae Stars

RR Lyrae variables are very old giant stars with radii about eight times that of the Sun and with surface temperatures of around 7000 K. Some of them pulsate in the fundamental mode and others pulsate in the first harmonic. Their periods range from a few hours to about a day. Their absolute magnitudes usually range about 0.5 and their brightness variations are from a quarter to two magnitudes.

Mira-Type Stars

Mira-type stars are red giants with surface temperatures around 3000 K. They are not entirely regular in their pulsations and have periods ranging from a month to

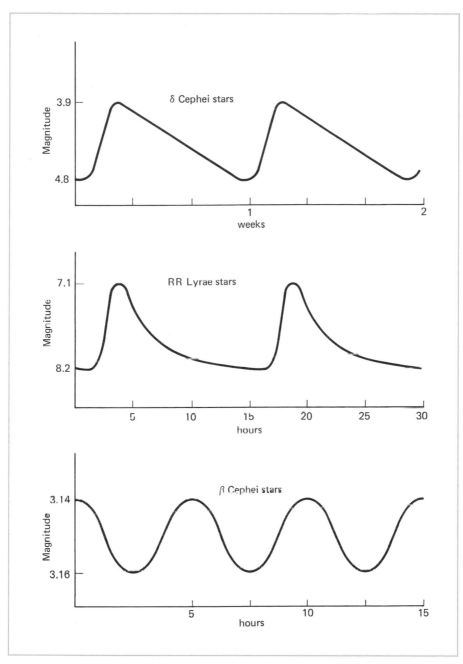

Figure 17.11 Some characteristic light curves of variable stars.

several years. They are very swollen stars, being some 100 times larger than the Sun with absolute magnitudes ranging around −1. About half of the variable red giant stars are of Mira type, the others having different light curves. Together they form the largest group of known variables.

Cepheid Variable Stars

These are very luminous (absolute magnitude around −0.5 to −6) giant stars, with diameters ranging from a little larger than our Sun's to 100 times as big. Their photospheric temperatures are similar to that of the Sun. The prototype, δ Cephei, varies from apparent magnitude 3.9 to 4.8 over a period of 5.4 days. Its absolute magnitude averages around −2.2.

Cepheid variable stars are very important to astronomers because they obey a *period-luminosity law* very well. This means that if a Cepheid variable can be identified by its light curve then its absolute magnitude can be fixed with a high degree of certainty. Knowing its apparent magnitude, its distance can then be calculated. If a Cepheid is known to be a member of a distant star system, then the distance to this system can be found. Since Cepheid variable stars are so luminous, they provide a valuable yardstick for measuring distances which are too great for parallax determinations to be successful.

This overview of variable stars has necessarily been brief and superficial, mentioning only the major types; there are many others. The study of these objects gives us an insight into the internal structure of stars and their evolution. The subject of stellar evolution is discussed in the next chapter.

17.8 Shells Around Stars

IRAS, the Infrared Astronomical Satellite, was a highly successful space probe. It made many discoveries. One of these was of the presence of large, cool, dust clouds – a sort of "infrared cirrus" – which pervade our Galaxy. I have already referred to the gas and dust between the stars but just how much there is was not known before the *IRAS* results.

Perhaps the most exciting of *IRAS*'s results is the detection of the large *infrared excess* of the star Vega (α Lyrae). As discussed in Chapter 15, stars behave (at least as far as their continuum emission is concerned) like black bodies. The emitted energy varies with wavelength in a very well-defined way. When *IRAS* was turned to Vega in 1983 the project scientists received a shock when they found that Vega was shining very brightly at infrared wavelengths, when it should be dim.

Further analysis revealed that Vega is surrounded by a shell of rocky material, perhaps amounting to several hundred times the mass of the Earth and spread around the star in a cloud, perhaps in a ring. The outer limits of the cloud are reckoned to orbit Vega at around 90 AU, which is roughly twice the radius of our Solar System. The particles of rock are thought to range about a millimetre on average, though probably with some rather larger. The average temperature of the shell is about 90 K, as determined by the strengths of the infrared emissions in different wavebands.

This result sparked a search of the databank of *IRAS* results and about forty or fifty other possible *circumstellar dust-shell* bearing candidates have been identified, though not all are as unambiguous as Vega.

Vega is a young star, perhaps about a thousand million years old. Many of the other stars with infrared excesses are also young. Inevitably this had lead to the suggestion that these circumstellar dust-shells might be embryonic planetary systems. We do not know. Recently there have been advances in the search for planets about other stars, but these are discussed in Chapter 19. The next chapter takes the story forward from the birth and early lives of stars to their "adulthood" and beyond.

Questions

1. Write an essay on the subject of the interstellar medium, detailing the natures and appearances of the various types of gaseous nebulae.

2. Outline the scenario that astronomers think gives rise to the birth of stars. Describe how the mass of a star determines its position on the main sequence as represented on a Hertzsprung–Russell diagram.

3. What are *open star clusters* and how do they relate to the formation of stars? Why are not all stars seen in the sky members of open clusters?

4. Distinguish between *intrinsic and extrinsic* variable stars, giving examples of each.

5. Write a short essay about variable stars.

6. Why has the satellite known as *IRAS* been so important in the investigation of the low temperature gas and dust in the Galaxy? Outline one other specific discovery made as a result of using *IRAS*.

Chapter 18

Stellar Evolution

T HIS chapter takes up the story of what is believed to happen to a star after it has settled onto the main sequence. Astronomers' ideas are still very theoretical, though there are plenty of examples of stars in apparently different stages of their lives to bolster the theorists' models. The timescales might be wrong, but astronomers do have confidence in the basic details.

18.1 The Main Sequence Phase

While a star is on the main sequence it is converting hydrogen to helium by thermonuclear reactions. For a star no more than about twice as massive as the Sun, the proton–proton cycle is the major energy-producing reaction. For heavier stars, which have higher central temperatures and pressures, the CNO cycle is dominant.

Just how long a star remains in its main-sequence, "hydrogen-burning" phase depends upon the mass of the star. A star of similar mass to the Sun will stay on the main sequence for around 10 000 million years. Those of smaller mass will last longer. Heavier stars will use up their hydrogen fuel at a faster rate. For example, we think that a star of 5 solar masses will stay on the main sequence for only 80 million years.

As a star ages, so the amount of helium in its core increases. The helium, being heavier than hydrogen, sinks to the centre of the core, leaving a hydrogen burning shell around it. These internal changes cause the star to brighten gradually as its surface swells and cools a little (the increase in the star's surface area more than compensates for the reduction in radiation flux that results from a slightly lower photospheric temperature).

Eventually, when about 20% of the star's hydrogen has been converted to helium, the star is no longer able to maintain the temperature necessary for nuclear reactions in its hydrogen burning shell. What happens next depends very much on the mass of the star.

18.2 The Fate of a Low-Mass Star

The term *low-mass star* denotes one which has a mass of less than half that of our Sun. When the hydrogen burning ceases, the star radically changes its structure. The helium core rapidly shrinks, pouring gravitational potential energy into the star's outer envelope. It is thought that much of this envelope is then puffed away into space, to form a *planetary nebula*.

Planetary nebulae take the form of shells of gaseous material surrounding the central star. An example of this phenomenon is the Ring Nebula, in the constellation of Lyra, shown in Fig. 18.1. It was discovered by Antoine Darquier in 1779. He compared its image in his small telescope to that of the planet Jupiter. Certainly it does appear about the same size, though vastly fainter. In a 6 inch (15 cm) reflector, or larger, it appears like an oval smoke ring. Larger telescopes are required to see its central star.

In one of his sky-sweeps in 1790 Sir William Herschel discovered the planetary nebula NGC 1514. It appeared to him as a faint star surrounded by a cloudy atmosphere. Herschel called it "A most singular phenomenon!" It was important to Herschel in that until then he had regarded "milky nebulosity" as being composed of innumerable stars too remote to be separately resolved. This object and others like it forced him to change his view. William Huggins added spectrographic confirmation in 1864. He found that planetary nebulae produce emission line spectra.

Figure 18.1 The Ring Nebula, M57, photographed by the author on 1988 July 19^d 00^h UT, using the 36 inch (0.9 m) Cassegrain reflector, formerly of the Royal Greenwich Observatory at their Herstmonceux site.

Astronomers now understand that the short-wave electromagnetic radiations from the central star cause the distended shell of gas to fluoresce. In typical examples the gas expands away from the central star with velocities from a few kilometres per second to a few tens of kilometres per second. As the gas expands so the nebula thins and eventually becomes invisible. For this reason all the planetary nebulae that can be seen have diameters not much more than a light year. Herschel (possibly influenced by Darquier's description of M57) gave planetary nebulae their name, as he found many that exhibit planet-like disks. However, many are very assymetric in appearance. For instance M27, the famous Dumbbell Nebula, certainly does not look anything like a planet (see Fig. 18.2).

While the outer layers of the star are being ejected, the bulk of its mass shrinks under gravitation, now that the nuclear fires are extinguished. Eventually the enormous pressures cause the boundaries of the atoms to break down, allowing the matter to reach incredible values of density.

A normal atom contains a tiny nucleus, with a diameter typically of around 1×10^{-14} m. The electrons "orbiting" the nucleus occupy a spherical region roughly 1×10^{-10} m across. Thus an atom is mostly empty space! If someone were to throw a brick at you, it might be difficult to imagine that most of what has hit you is a vacuum but this is, in fact, true! Of course, in the high-temperature environment

Figure 18.2 The Dumbbell Nebula. Photograph taken at the prime focus of the Isaac Newton Telescope, formerly at Herstmonceux. RGO photograph.

inside a star atoms cannot survive intact. The electrons are stripped from the nuclei to form a plasma of interspersed electrons and nuclei.

Under the enormous pressure produced by its own weight, the dead star's electrons and nuclei are forced together very much more closely than is possible even in normal atoms. In terms of quantum mechanics, the electrons occupy energy levels inside the star, rather than relative to atomic nuclei. Moreover, in a white dwarf all the lowest energy levels are filled up with few or no spaces left. Physicists say that matter in this condition is *electron-degenerate*. The collapse of the dead star is then halted by the pressures caused by this quantum-mechanical limitation, which we call *electron degeneracy pressure*.

The final state of the star, where most of it is made of electron-degenerate material, is called a *white dwarf*. Its radius is about one hundredth that of the Sun (little bigger than the Earth) and its density is around 3×10^8 kg/m^3, 300 000 times that of water. In fact, the more massive white dwarfs have smaller radii and so even higher densities. They are peculiar objects. Their central temperatures are only around 1 million K and this temperature varies little throughout the major part of their interiors. We cannot think of white dwarfs as being gaseous. They are more like huge crystals, gradually cooling. Only quantum physics can adequately describe matter in this state. Their surfaces are very hot, around 100 000 K for newly formed white dwarfs. They are thought to rotate very fast, taking perhaps only 10 seconds to turn once on their axes.

It is also believed that the magnetic field from the original star is concentrated to a strength millions of times greater than that of our Sun's field. The transition from a low mass main sequence star to a white dwarf takes only a few thousand years, and so astronomers have only a sketchy idea of the path such a star takes across the H–R diagram. However, over a hundred of the final white dwarfs have been discovered.

The star Sirius has the greatest apparent brightness of any star in the sky. It is the leading star in the constellation of Canis Major. It lies south of the celestial equator, but from mid-northern latitudes it can be seen low down in the sky during winter evenings. Sirius is fairly close to us, at 2.6 parsecs, and this is the reason for its apparent brilliance. In fact Sirius is a main sequence star of spectral type A1 and of mass twice that of the Sun.

After studying its proper motion Friedrich Bessel announced, in 1834, that Sirius has some unseen companion. He came to this conclusion because of an unaccountable "wobble" in the path of the star (produced by the orbital motions of Sirius around the barycentre of the binary system). However, no obvious companion could be seen. The telescope maker Alvan Clarke found it in 1862, when he was testing out a large refractor. This discovery immediately set astronomers a problem. The motions of the two stars showed that the companion had a mass fully half that of Sirius, but it was only one-ten-thousandth as bright. It was of similar colour to Sirius and so must have a similar surface temperature. These observations could only be reconciled if the companion had a radius little bigger than the Earth. It would then have an incredible density. Thus was discovered the first example of a white dwarf.

The central stars we see in many planetary nebulae are white dwarfs. In fact, white dwarfs must be common. Astronomers estimate that about 10% of the stars in our Galaxy are these stellar relics, though their low luminosities make them difficult to detect. Over the course of aeons white dwarfs gradually cool, eventually disappearing from sight as they radiate the last remnants of their thermal energy into space. The central star of the Ring Nebula, shown in Fig. 18.1 is very blue (and therefore very hot) and is thought to be on the verge of becoming a white dwarf.

Interestingly, there is considerable disagreement between amateur astronomers about the size of telescope needed to show M57's central star. My personal opinion is that it might well be variable in its brightness (I am certainly not the first to think this), undergoing eruptions while it progressively rearranges itself into a white dwarf. There is just a chance that, quite by accident, I captured it during an outburst when I photographed the nebula on 19 July 1988 at 00 ^{h} UT. The central star does certainly seem more prominent than usual compared with the other stars shown, while those that are normally of similar brightness to it are too faint to be seen on my photograph (see Fig. 8.1).

18.3 The Evolution of a Star Like the Sun

Considering stars of mass between half and twice that of the Sun, their nuclear reactions can go beyond the stage of converting hydrogen to helium. As with the less massive stars, a helium core develops with the "burning" hydrogen confined to a shell around the core. When the nuclear reactions eventually dwindle the core contracts and becomes electron-degenerate as before. The consequent release of gravitational potential energy causes the outer regions of the star to swell up and redden. The star will have become a red giant.

When this happens to our Sun, in about 5000 million years' time, it will attain a radius about 30 to 50 times its present size and a surface temperature of about 3500 K. Some popular books suggest that at this stage the Sun will swallow up the orbits of the inner planets, including the Earth. This is not so, although Mercury might well be engulfed. However, it is true to say that the Sun's luminosity will increase to around 500 to 1000 times its present value. It will probably sustain that for a few hundred million years. As a result the Earth will be stripped of its atmosphere and the oceans will boil away, leaving the Earth a scorched and dead globe.

The degenerate helium cores of these stars can contract under their own weight until they reach higher temperatures than those in their less massive counterparts. When the temperature reaches about 100 000 K the fusion of helium to make carbon and oxygen is possible. However, before these reactions can begin the core has to expand to remove (in its outer parts at least) degeneracy. This only happens at a higher temperature. When that higher temperature is reached, helium "burning" begins violently. This produces a surge in luminosity lasting a few centuries, known as the *helium flash*.

It is at this time that a star such as our Sun will give one final heave and puff out its tenuous outer layers so that, this time, all the inner planets are indeed engulfed. As this happens a star like our Sun will lose a large fraction of its mass, which will probably go to form a planetary nebula.

After the helium flash the star will settle into a new equilibrium condition. Around the inert core will lie a zone of helium burning, and above this a zone of

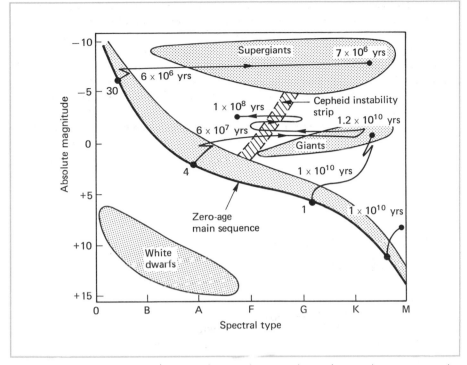

Figure 18.3 Evolutionary tracks across the H–R diagram. The numbers on the positions on the zero-age main sequence correspond to the masses of the stars. The ages of the stars as they leave the main sequence and as they attain the "helium-burning" stage are shown.

hydrogen burning. The outer layers will gradually expand a little further and the overall luminosity of the star will increase to over a thousand times its main sequence brightness. The time-scales of these events are a little uncertain, as are the precise details such as the amount of mass lost by ejection.

Eventually the helium and hydrogen fusion reactions die. The core of the star then evolves to become a white dwarf, the outer envelope having been lost to form a planetary nebula, as already mentioned.

The star's changes in luminosity and effective temperature as it evolves can be plotted as a track across the H–R diagram (see Fig. 18.3). As the star changes from being a red giant to a white dwarf it passes through the Cepheid instability strip. For a while the star then becomes a Mira-type variable. Astronomers think it likely that the final white dwarf will have only two-thirds of the initial mass of the star, the rest being lost through stellar winds and planetary nebula forming outbursts.

18.4 The Evolution of a Massive Star

Stars of mass greater than about 2.2 times that of the Sun but less than 5 solar masses do not develop largely degenerate cores. In these stars the helium-burning phase begins steadily and so they do not show the helium flash phenomenon.

Instead, they exhibit complex looping paths across the red giant region of the HR diagram as they undergo successive stages of core and envelope contraction and expansion. Eventually they shed a great deal of their initial mass in repeated outbursts (forming planetary nebulae) and the stellar remains become white dwarfs.

Astrophysicists are at their most uncertain when they try and predict the fates of stars of still greater mass. The amount of mass loss is crucial in determining what happens to a star, and this is a very poorly understood factor. Certainly, massive stars use up their fuel reserves voraciously. A star of 30 solar masses will remain on the main sequence for only 6 million years. During this stage the star is a giant of spectral type O, with a luminosity of more than 100 000 times that of our Sun and a radius 7 times greater. At the end of its main-sequence life, the core of the star contracts and the envelope expands as a helium-burning shell ignites.

The subsequent path of the star across the H–R diagram is much simpler, being virtually a straight line with no loops. The star passes through the Cepheid instability strip, for a time becoming a variable. About a million years after it left the main sequence the star has increased its diameter from its initial 7 times that of our Sun to around 1000 Sun-diameters. The star has then become a red supergiant. At this stage the star has an inert core rich in carbon, nitrogen and oxygen. The helium-burning zone envelops the core, in turn overlaid by the hydrogen-burning zone, then the outer layers of the star.

As the reaction rates in the hydrogen and helium zones dwindle, so the core shrinks and heats even further. Nuclear reactions begin in the core, synthesising progressively heavier elements, such as magnesium, sodium, aluminium and neon. As each burning stage begins, so the core shrinks and heats up even more. Iron ultimately builds up in the core of the star, being synthesised from the fusion of silicon, and this leads to the star's downfall.

The fusion of the lighter elements into heavier ones result in the release of energy but to get iron to build up more massive nuclei energy has to be put in. Once formed, the iron remains inert. The iron core contracts until its temperature reaches 5000 million K, when photons of electromagnetic radiation have energies sufficient to begin breaking the iron down into less massive elements, such as helium. As a result the core very rapidly loses thermal energy, triggering a catastrophic collapse. This takes only a few seconds and the resulting implosion of the star's outer layers and the sudden release of energy-carrying particles, called neutrinos, cause runaway nuclear reactions in these outer layers.

The result is a tremendous explosion, a *supernova*. Most of the star's material is blasted away into space in a spectacular outburst, but the kickback on the star's iron core causes it to collapse to a super-dense object known as a *neutron star*. A representation of the structure of a massive star at the onset of a supernova is given in Fig. 18.4. Supernovae are obviously very important events and are discussed in more detail in the Chapter 20. Meanwhile, we now consider neutron stars in more detail.

18.5 Neutron Stars

White dwarfs can exist only if their masses are less than 1.4 times the mass of our Sun. This is the *Chandresekhar limit*, so named after the scientist who theor-

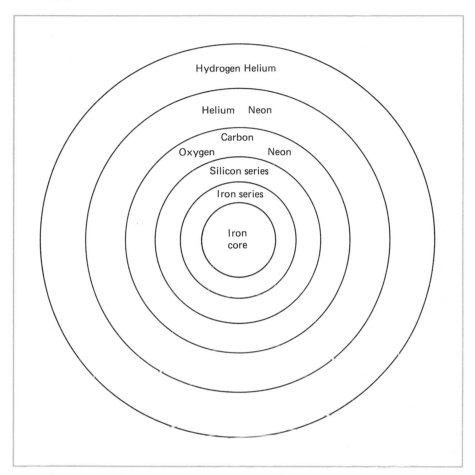

Figure 18.4 The structure of a massive star just before the onset of a supernova.

etically demonstrated its existance. If the dead star is of still greater mass then electron degeneracy pressure cannot withstand the crushing forces of the star's great weight.

Such a stellar relic would continue to collapse until electrons "tunnel" into protons, creating neutrons. The resulting object would consist almost entirely of neutrons. It would be a *neutron star*, with a diameter of less than 20 km and a central density of a million million million kilograms per cubic metre! Needless to say, physicists have no experience of how matter behaves under such extreme conditions but they have attempted to construct theoretical models of the structure of these bizarre objects. It seems that the major part of a neutron star's body is made up of a fluid of neutrons, overlaid by a crust of solid neutron material.

Neutron stars contain much of the angular momentum of the original star. The precursor, though rotating slowly, had a lot of angular momentum owing to its great size. As a consequence of its shrinkage to a very small size, the neutron star must rotate very fast. Newly formed neutron stars probably spin on their axes approximately 1000 times every second. However, an object on the surface of a neutron star would not be flung off into space. The gravitational pull on its surface is over 200 million million times stronger than on the surface of the Earth! The magnetic field of the original star is similarly concentrated. It has a value about a million million times stronger than the Earth's field.

This incredibly powerful magnetic field must affect the photosphere of the neutron star, though the photosphere is only a few centimetres thick because of the intense gravitational pull. Theorists think that the temperature is around 10 million K for most of a neutron star's bulk. Much of the radiation from the star, as well as electrified particles, must escape from the regions of the magnetic poles, where the lines of magnetic force enter the photosphere nearly perpendicular to it. Also, the magnetic and rotation axes may be inclined to each other. This would give rise to a "flashing" effect as the beam of a rotating neutron star sweeps across our line of sight. These flashes have actually been observed.

At the Mullard Radio Observatory of the University of Cambridge, England, in 1967 a post-graduate student Jocelyn Bell (now Burnell) was engaged in a programme of mapping the sky at radio wavelengths as part of her work for her PhD thesis. She came across a signal that she at first took to be artificial and of terrestrial origin. Continued observations showed that the object moved with the stars, placing it beyond the Solar System. The object produced radio pulses every 1.337 seconds. In the years that followed other examples of pulsating radio objects, or *pulsars*, were found. They have periods ranging from 0.006 second to 4.3 seconds.

Theory soon linked these radio objects to neutron stars, and they are now thought to be one and the same. By Earthly standards these objects are very good timekeepers, but precise measurements show that they are all very gradually slowing down. The average rate of slowing seems to be proportional to the reciprocal of the period of the pulsar and ranges from 10 millionths of a second per year to 10 million millionths of a second per year. This may not sound very much but it represents a considerable removal of energy from the star. The fact that astronomers have found no pulsars of very long period must mean that the beaming mechanism shuts off 10 million, or so, years after the pulsar has first formed. It is not known why it does this or how the beaming mechanism works.

18.6 The Evolution of Binary Star Systems

If the two stars that form a binary system are far apart from each other then they will evolve independently. The rates of evolution and the way each star evolves will be determined solely by their masses, as is the case for single stars. On the other hand, if the stars orbit close to each other there may well be a net transfer of mass from one star to the other. The evolution of each star then becomes much more complicated.

Some way between the two stars lies a position, known as a *Langrangian point*, at which any object placed there would experience equal gravitational forces from each star. An object exactly at this position would not move with respect to either star. However, if it was slightly displaced towards one star then the gravitational pull from that star would win and the object would fall towards it. Hence the Langrangian point is a position of unstable equilibrium.

The Langrangian point lies at the intersection of the gravitational equipotential surfaces of the two stars. These are imaginary surfaces where any objects on one of these surfaces would have common values of gravitational potential energy. The equipotential surfaces form egg-shaped regions around the stars, called *Roche lobes*. The Langrangian point is at the "sharp end" of the eggs, where they intersect. The best way of thinking about these Roche lobes is to imagine them as "regions of influence" of the gravitational fields of the stars. If a star, at any time in its career, expands to fill its own Roche lobe then material from the star will flow through the Langrangian point and on to the other star (see Fig. 18.5).

As the material from one star falls towards the other a great deal of its gravitational potential energy is released, so giving rise to energetic phenomena. A large portion of the X-ray sources in the sky are the result of interacting binary stars. In these cases the mass from the donor star falls onto a compact object, such as a white dwarf or a neutron star, which have powerfully concentrated gravitational fields.

If the donor star is much more massive than the other, the material falls straight onto it. If the two stars are of similar mass, the transferred material tends to form a ring of material, an *accretion disk*, around the receptor star. When the fall of the stellar material is arrested, either at the surface or at the accretion disk of the second star, it produces a region of very high temperature – often around 10 million K. These high temperatures result in the emission of substantial quantities of X-ray radiation.

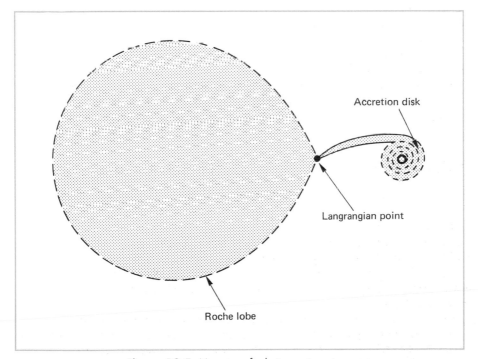

Figure 18.5 Mass transfer between two stars.

A general property of X-ray sources is that they are variable. Most are also extremely powerful. Their X-ray luminosities can be tens of thousands of times greater than the total output of the Sun, summed over all wavelengths. Some show periodic variations. Astronomers think that these are due to the effects of the orbital motions of the stars, or the rotation of the compact star (where this star has localised "hot spots" on its surface), or the precession of an accretion disk. Other stars show much more erratic and vigorous changes in X-ray luminosity. These are termed *X-ray bursters*. At present astronomers are at a loss to explain their behaviour.

Even more strange are the *γ-ray bursters*, as these appear to be single compact sources. At the moment γ-ray bursters are very much a mystery. We do not even know how far away they are. They could be very remote, perhaps out at the edge of the visible universe, in which case they are unbelievably powerful. If they are more local then they are still enigmatic, because astronomers have not succeeded in definitely identifying them with any known objects. Researchers' results and the conclusions they draw from them are at the present conflicting.

It is very likely that *Wolf-Rayet stars* (see Chapter 16) are the result of mass transfer processes in binary systems. Nearly 200 examples are at present known. They exhibit peculiar spectra, dominated by bright emission lines. The emission lines belong to such elements as helium, nitrogen and carbon. These lines are highly broadened, indicating that the gases causing them are being shed from the star at velocities of several thousand kilometres per second. It certainly appears that much of the star's hydrogen envelope has been lost, laying bare the helium-rich interior. It could be that the star has reached the red giant stage in its life, filled its Roche lobe, and then had its outer envelope stripped away by a companion star.

Questions

1 Describe the evolution of a star of similar mass to the Sun. Sketch the path it describes across the Hertzsprung–Russell diagram as it evolves.

2 Write an essay about *red giant* stars.

3 Describe the natures of *white dwarf* stars. How do these objects relate to stellar evolution?

4 Write an essay on the subject of the evolution of massive stars (those of more than ten solar masses).

5 Describe the ways in which stars can interact if they are members of binary systems.

6 Write an essay about *neutron stars*.

7 What are *planetary nebulae* and how are they formed?

8 A *white dwarf* star, a *red giant* star, a star similar to our *Sun* at present, and a *planetary nebula* are all very different types of object. Describe them and explain how astronomers think they are related to one another.

Chapter 19

Brown Dwarfs and Extra-solar Planets

E ARLIER in this book we took a look at how astronomers think the Solar System was formed from a rotating cloudlet that span into a disk, with the Sun formed at its central hub while the planets formed from rings of matter in the orbiting disk.

Since other stars are also formed from similar rotating cloudlets, you might have wondered whether planetary systems perhaps surround other stars. Most astronomers also wondered this and many worked hard to find out. In time some positive evidence began to accumulate, but the really indisputable proof came in late 1995. We now know for sure that our Solar System is not unique. At least some other stars do, indeed, act as parents to their own families of planets.

19.1 Early Mistakes

Until recently, planets were held to have been detected around certain stars, particularly Barnard's Star and 61 Cygni. The method used was to take accurate measures of the slight periodic "wobbles" that any massive planets would make as they orbited the parent star. Figure 19.1 illustrates the principle. In fact, as the diagram shows, the "wobble" would be superimposed on the star's own proper motion through space and the result would be that the star would describe a slightly wavy line, rather than the straight line that would be due to pure proper motion.

The method was perfectly valid in principle. However, the trouble was that the amplitude of the oscillations would be tiny – in fact right at the limit of detection and measurement. In the last decade or so further work has been done, using refined techniques, and the early results have now been disproved. The astrometrists carrying out the original work had got it wrong. Neither Barnard's Star, 61 Cygni nor any of the others apparently found before the 1980s has an associated planetary system.

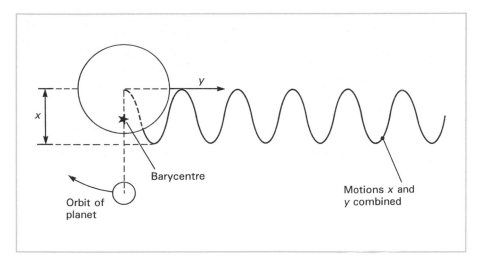

Figure 19.1 The presence of an orbiting planet causes the centre of the star to move around the barycentre, producing the apparent lateral to-and-from shift of size x. Added to its steady and linear proper motion, y, the result is that the star appears to trace out the wavy line shown, against the backdrop of more distant stars.

The mistakes certainly weren't confined to the years before the 1980s. Speckle interferometry (see Chapter 4) had apparently revealed a sub-stellar companion to the star called Van Biesbroeck 8. However, after the initial "discovery" all further observations failed to find it. The companion had vanished! Of course, the answer is that it was never really there in the first place.

The other notable false alarm of later years was made using a technique that is relevant to the story of the later successes and so we will discuss it more fully here. Andrew Lyne, a radio astronomer at the University of Manchester, had been conducting research on pulsars at the famous Jodrell Bank installation in Chesire. An expert on pulsars, he has discovered more of them than anybody else in the world, to date. Pulsars (neutron stars) are discussed more fully earlier in this book, but of relevance here is the fact that they are extremely accurate timekeepers. Apart from the occasional sudden anomaly (caused by a structural rearrangement), their pulses arrive at very steady and predictable rates.

One of the pulsars Professor Lyne investigated in the mid-1990s proved to rather different. The pulses did not arrive at a set frequency. In fact the pulse frequency varied very slightly in a repeating cycle of six months' period. This could be explained if the pulsar was cyclically moving to and fro around its mean distance from us. When the pulsar is approaching the train of pulses emitted by it would be compressed and so arrive at a slightly higher frequency. When receding the train of pulses would be stretched out and the arrival frequency would be slightly lower. You will recognise that this is simply the Doppler effect in action, and the situation is illustrated in Fig. 19.2 (*overleaf*).

Lyne announced his results, which received a large amount of publicity. Unfortunately it was only afterwards that he realised that a computer program had not properly corrected for the Doppler shift due to the Earth's own motions (all the received pulses of neutron stars are effected by this, to a varying degree

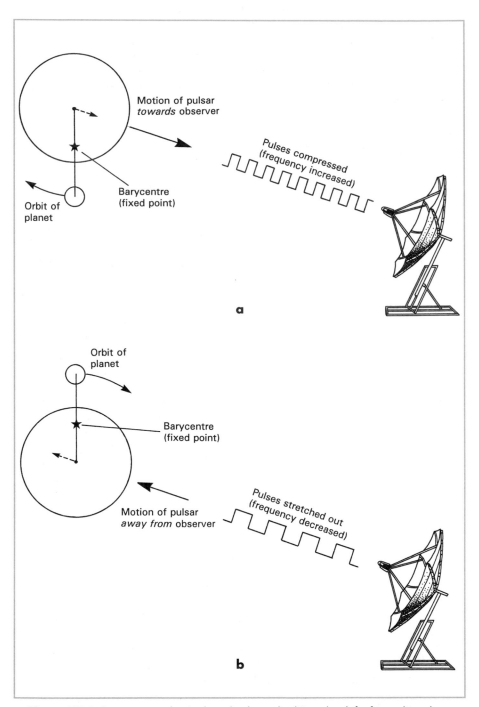

Figure 19.2 Detecting a pulsar's planet by the cyclical Doppler shift of its radio pulses.

depending on their positions on the sky). To his horror he found that when he corrected the mistake, the pulsar variations vanished. There was no planet orbiting the pulsar after all!

So much for the mistakes. Lessons were learned and techniques were refined. Now we *do* have evidence for extra-solar planetary systems, or at the very least evidence for orbiting objects which are not themselves true stars. The first of these was discovered in 1988.

9.2 HD 114762

Once again the principle of detecting a cyclical Doppler shift was used. However, this time the measurements were made in the visible part of the spectrum. The star with the decidedly unromantic designation of HD 114762 is a cool main sequence star (spectral type F9$_v$), about 140 light years away, in the constellation of Coma Berenices.

Very high dispersion spectra of this star were taken by a team headed by D. W. Latham for quite another purpose. They used a 1.5 m reflector in Massachusetts and discovered the periodic spectral line shifts quite serendipitously. Other teams have confirmed the presence of a large non-stellar body orbiting HD 114762 and refined the earlier determinations of its mass and orbital details.

It now appears that the body orbits the star at a mean distance of 0.34 AU with a period of 89 days. The orbit does seem to be highly eccentric. As far as determining its mass goes, that is a little more problematical, owing to what I call "the angle-of-dangle problem".

The essential point is that we can only detect radial motions from Doppler shifts. That is, only the component of a star's motion which is along our line of sight. As Fig. 19.3 (*overleaf*) illustrates, a given Doppler shift might result from a relatively small planet orbiting close to our line of sight. Alternatively, it might be due to a much more massive object whose orbital plane is presented to us at a much greater angle. Therefore the best we can do at present is to put a lower mass limit to the object orbiting HD 114762. It turns out to be about 9 times the mass of our planet Jupiter.

Is this body a true planet? If it has a mass of just 9 Jupiters then this seems likely. However, if the mass is much greater then we could perhaps better regard it as a *brown dwarf* – a star of insufficient mass to create a high enough core temperature for it to sustain (or even to ignite at all) its nuclear furnace after it first formed. More than about 80 Jupiter-masses of material would be required to produce a true star. In effect, a brown dwarf is a failed star.

What distinguishes a planet from a brown dwarf? Well, if the mass of a body is more than 10 to 20 Jupiters then astronomers would prefer to call it a brown dwarf. However, the dividing line in terms of mass is blurred to the extreme, and is rather more a question of semantics. Perhaps a better distinction lies in the way in which the object formed. If it formed at the central hub of a rotating cloudlet, then it should be considered to be star-like and so the term brown dwarf might be more appropriate. If it formed in the rotating disk then it would be called a planet. However, even this doesn't provide a cast-iron definition. (Consider two closely

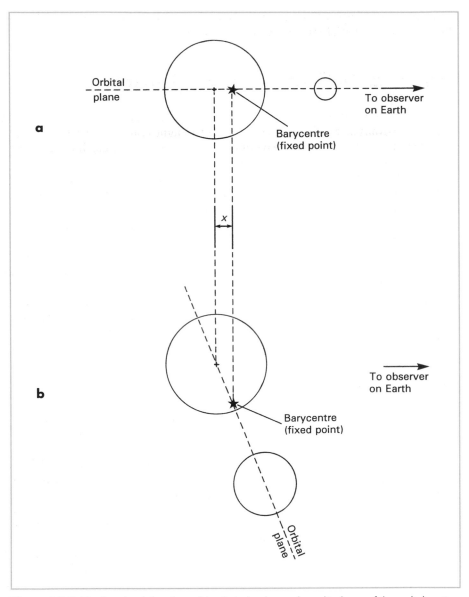

Figure 19.3 The "angle of dangle problem". Is the detected amplitude, x, of the radial motion caused by a body of small mass orbiting in nearly the same plane that we see it (as in **a**), or a much larger body whose orbital plane is presented to us as in **b**?

revolving binary stars – is one of them, then, a planet?). At this time we certainly cannot tell whether any massive orbiting body has actually formed from its own embryonic cloudlet, or as a satellite of its parent star.

   To summarise, if the mass of any orbiting body is less than 10 to 20 times Jupiter's mass, then astronomers regard it as a planet. More massive, and the orbiting body is regarded as a brown dwarf. The body around HD 114 762 might be a planet, but it might instead be a brown dwarf.

19.3 A Definite Brown Dwarf

One body that is certainly too massive to be regarded as a planet is that orbiting an M-type dwarf star called Gliese 229, about 19 light years from us. Gliese 229 is of the eigth magnitude and appears in the constellation of Lepus, though no ordinary equipment will show its companion.

The companion body was first identified by means of a stellar coronograph (the analogue of the solar coronograph but designed for use on the stars) with the 1.5 m and later the 5 m reflectors on Mount Palomar in 1995. The images the coronograph gave were sharpened by the use of adaptive optics. Tadashi Nakajima and his team were able to directly image the body. It appears to have a mass of around 40 times that of Jupiter (with a high degree of uncertainty in this figure) and a photospheric temperature of no more than 1000 K. Its orbit is still highly uncertain.

There are a couple of other possible brown dwarf candidates, both in the Pleiades star cluster, but Gliese 229 B, as it is known, is the only one astronomers have confidence in regarding as a correct identification at present.

19.4 PSR 1957+12

About the time that Andrew Lyne was working at Jodrell Bank producing his unfortunately erroneous result another radio astronomer, Alex Wolszczan of the Pennsylvania State University, also found a pulsar, about 300 parsecs away in Virgo, which emitted a cyclically shifting stream of pulses in the same sort of way. Wolszczan had used the Arecibo radio telescope to survey a small strip of sky above it. However, this time the signature was much more complicated.

Three companions seem to orbit the pulsar. The innermost takes 23 days to go round at 0.19 AU from the pulsar. The second takes 66 days to orbit at 0.36 AU and the third takes 95 days to orbit at its distance of 0.47 AU. The three planets move in such a way that they come fairly close together every 200 days or so, and their mutual gravitational pulls disturb each of their orbits in a very predictable, if complicated, manner. Subsequent observations revealed that the orbits and periods of the planets are affected in just the way predicted. It seems that this time there is no doubt. At least three planetary bodies do, indeed, orbit the pulsar PSR 1957+12.

In increasing order of distance from the pulsar, the calculated masses of the planets are: 0.000 05 Jupiter (0.16 times the mass of the Earth), 0.011 Jupiter (3.5 Earth-masses) and 0.009 Jupiter (2.9 Earth-masses). Of course, these are the lower limits. Remember the effect of the "angle of dangle problem"!

It must be admitted that a pulsar (neutron star) is just about the last place astronomers expected to find orbiting planets, reasoning that any planetary system ought to have been destroyed by the later stages of the life of the progenitor star and the supernova explosion that led to the formation of the pulsar. Perhaps these planets were formed the debris of the exploding star? This is mere speculation – at the moment nobody knows.

19.5 51 Pegasi

Didier Queloz and Michel Mayor, two Swiss astronomers working at Haute Provence Observatory, in France, found evidence for a non-stellar companion orbiting the star 51 Pegasi. By examining high-dispersion spectra of the star they uncovered a cyclic to-and-from shift in the spectrum – a planet had been revealed to orbit this solar-type ($G2_v$) star.

By the spring of 1995 they had accumulated enough data to be reasonably confident, but they had to wait until the summer before they could continue their observations as 51 Pegasi moved near to conjuction with the Sun. When they could begin again, they found that the radial velocity curve of the star exactly fitted their prediction. They could then be absolutely sure and they announced their discovery in an astronomical meeting in Florence in October 1995. This was the first planet to be discovered around a star that is similar to the Sun (indeed, 51 Pegasi is the Sun's twin).

However, the planet itself is far from anything that was expected. It seems to have a mass equivalent to about 0.6 to 2 Jupiters and it orbits 51 Pegasi with a period of just over 4 days and at a distance of only about 7 million kilometres (about 0.05 AU) from the star! What a strange, incredibly hot world this must be.

19.6 47 Ursae Majoris

Geoffrey Marcy and Paul Butler head a small team of astronomers that have been using the 3 m Shane reflector at Lick Observatory to search for extra-solar planets. They also had been looking for the tiny Doppler shifts they would produce in very high-dispersion spectra of the parent stars. However, their reduction procedures were not as rapid as those of the Swiss team.

Marcy's group had been in operation for about ten years when Mayor and Queloz announced their discovery. The team at Lick rapidly confirmed this discovery and soon afterwards they announced new discoveries of their own. One of these is of a 2.3 Jupiter-mass (lower limit, as always) body orbiting 47 Ursae Majoris at a distance of about 2.1 AU and with a period of about three years.

The parent star is of spectral and luminosity classification $G0_v$, which makes it broadly similar to our Sun. Actually it was the policy of Marcy's team that they deliberately select solar-type stars for their invesigations. One is tempted to wonder if, at its distance of 46 light years from us, lies a planetary system something like our own? Of course, all such flights of fancy are without any real justification at present.

19.7 70 Virginis

Another find by the Lick Observatory team, this time of a body of 6.5 Jupiters minimum mass which takes four months to orbit the star 70 Virginis, at 0.43 AU from it.

This star is also not too unlike our own Sun, having a classification of $G0_v$. It is situated at about 80 light years from us. Calculations suggest that the temperature of the body ought to be around 80 °C, but the various claims that liquid water must exist on it are sheer fantasy. Water may indeed exist there, and equally it may not. We most definitely do not know what the conditions might be on it, if indeed this body is even a true planet and not a brown dwarf.

19.8 Rho Cancri

This is yet another discovery by the team at the Lick Observatory. In this case the parent star is of spectral type $G8_v$ and lies at about 46 light years from us. The lower mass limit of the companion has been determined as 0.9 Jupiters and it must orbit the star at only about 0.1 AU with a period of 14 days. As always, the 'angle of dangle problem' might mean that the companion is really a brown dwarf.

19.9 Lalande 21185

Using a special (and elaborate) photometric device on the 30 inch (76 cm) refractor at the Allegheny Observatory, a team of amateur and professionals led by George Gateshead has made the provisional discovery that the M2 dwarf star called Lalande 21185, in the constellation of Ursa Major, has two detectable bodies in orbit around it. The technique they used pins down the orbital data of the bodies more precisely than those using the Doppler effect, and the bodies do seem to have masses of just under and just over Jupiter's. They orbit the star at 2.2 AU and 11 AU, respectively. However, these first results *are* provisional at present.

The detection of extra-solar planets is a new and exciting event in astronomy. I must emphasis that the details I have given here are just what is known at the time of writing (the summer of 1996). One can expect further developments. Some of the details given here may well be revised even before this book appears in print. Certainly I would be surprised if new planetary candidates are not discovered before then.

Naturally, the discovery of other planetary systems excites speculation about extraterrestrial life. However, I should emphasise that that is what it is at present – mere speculation. Many people think that the Universe should be teeming with life. Others insist that we are alone in the Cosmos. Perhaps the truth lies somewhere between. As yet we do not know. However, what we can say is that we have made the first tiny step in answering this question – the detection of planets orbiting other suns.

Questions

1 Write an essay about the methods used to probe stars for possible signs of associated planetary systems.

2 What is a *brown dwarf*?

3 Outline the details of six candidates for possible extra-solar planetary sytems.

Chapter 20

Violent Stars

GAZING up at the starlit sky on most evenings, we feel that the heavens are unchanging and serenely peaceful. Just once in a while we may see something to remind us that this impression is far from true.

20.1 Novae

As twilight fell on the evening of 29 August 1975 I happened to step outside and glance up at the heavens. A few of the brightest stars were becoming visible against the deep-blue sky. Arcturus and Vega were prominent and I could see several of the brightest stars in Cygnus – but something was wrong. Cygnus did not look quite right. Then I realised – Cygnus had gained a bright star to the north-east of Deneb! I could hardly believe it – what had I found? Was it a distant supernova? Maybe it was a less powerful, and so closer, *nova* – an erupting star. I telephoned an amateur astronomer friend and he confirmed the new star (and my sanity!).

Next, I telephoned the Royal Greenwich Observatory, then at their Sussex home of Herstmonceux. An astronomer there told me that they were about to turn the Isaac Newton Telescope (which has since been moved to La Palma) onto the object. It was a nova and had been discovered several hours earlier, before darkness fell over Europe, by several Japanese amateur astronomers. I was one of hundreds to find the nova independently that first night.

The original star is not shown on pre-discovery photographs and so was definitely fainter than the twentieth magnitude. When I saw the object it was of the second magnitude and it reached maximum brightness on the following night, then being at magnitude 1.8. Hence the original star had brightened by a factor of several tens of millions in a time period of only a day or two! Subsequent nights showed the nova in rapid decline. Within a fortnight it had become too faint to see with the naked eye and Cygnus had once more resumed its normal appearance.

About thirty novae appear in our Galaxy each year, though they rarely become bright enough to be visible to the naked eye. In order to explain them astronomers once again invoke mass transfer processes between a binary pair of stars (see Chapter 18). What happens is that hydrogen gas from the donor star piles onto the surface of a white dwarf and eventually triggers an explosive nuclear reaction which blows away a portion of its outer layers. The loss of mass in a nova outburst amounts to about one-ten-thousandth of the mass of the initial star. The sites of many novae show rapidly expanding nebulosity after the outburst. The nova shines with an intrinsic luminosity up to a million times greater than our Sun, taking only a few days to reach this peak.

Nova Cygni 1975 was notable for its great intrinsic luminosity (about a million times that of the Sun) and its very rapid rise to maximum brightness and its subsequent rapid fall. Others are much slower, like Nova Vulpeculae 1979 shown in Fig. 20.1.

Some novae result in a star increasing its output by a factor of 10 to 50 times. These are termed *dwarf novae*. Variable star observers sometimes call dwarf novae U Geminorum or Z Camelopardalis stars, because these were the first stars found which show this behaviour. Many dwarf novae give repeat performances every few months. The brighter novae are thought to erupt once in every few tens of thousands of years, on average. The star T Coronae Borealis is one example of novae intermediate in terms of brightness and period between outbursts. It is normally of the tenth magnitude, but rose to the second magnitude for a brief spell in 1866 and again in 1946. Astronomers call these *recurrent novae*.

20.2 Supernovae

A supernova is a colossal stellar explosion, where the star literally blows itself to pieces. On 4 July 1054 Chinese astronomers witnessed the appearance of a bright star in the constellation of Taurus. In the few surviving reports from the time, the brightness of the object is compared to that of the planet Venus. These reports state that the object was pinkish in colour and that it could be seen in broad daylight (as Venus can when the sky is really clear) for three weeks. Twenty-one months passed before the object faded to below naked-eye visibility.

In 1731 the English amateur astronomer John Bevis found a dim nebula in the constellation of Taurus. The French comet hunter Charles Messier rediscovered the object in 1758. He thought he had found Halley's comet, which was predicted to make an appearence in that year. Historians say that it was this error which prompted Messier to draw up his list of objects which might be mistaken for a comet. Certainly the nebula in Taurus is number 1 in Messier's catalogue. It is now known that the bright star reported by the Chinese was a supernova and the nebula M1 is the wreck of the old star, or *supernova remnant*.

Lord Rosse's 72 inch (1.8 m) aperture reflector revealed the nebula's complex filamentary structure in 1845. This structure led to the object being nicknamed the Crab Nebula, a name which we still use today. The 1054 supernova was not the first to be observed, nor was it the last. Astronomers estimate that on average one supernova occurs in our Galaxy every fifty years, though most are very remote and obscured by interstellar matter.

Supernovae appear to fall into two distinct groups. *Type I* supernovae are thought to be the result of mass transfer processes between a binary pair of stars, where one of the stars is a white dwarf. Like novae, matter is transferred from a

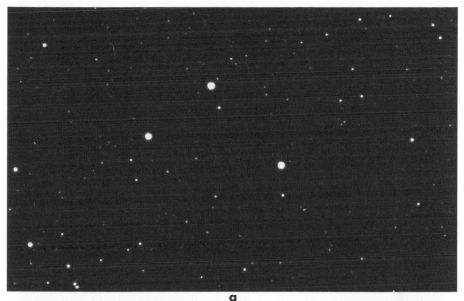

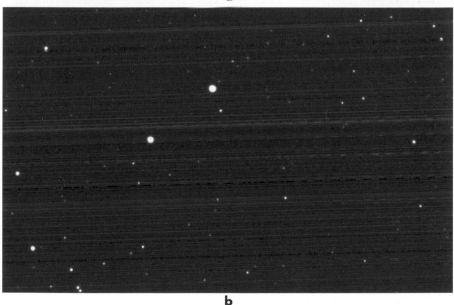

Figure 20.1 Nova Vulpeculae 1979; this shows **a** the nova on 20 July and **b** on 22 August, when it had dimmed considerably. Unlike Nova Cygni 1975, this nova was very slow to rise and fall in brightness. Photographs taken with the 13 inch (330 mm) astrographic refractor at Herstmonceux. RGO photographs.

swollen star onto a white dwarf, via an accretion disk. If the white dwarf collects enough matter for its mass to exceed 1.4 times the mass of the Sun, it cannot remain as a white dwarf. It must collapse to form a neutron star. This collapse is extremely rapid, leading to colossal outpourings of energy and runaway nuclear reactions in the outer zones of the star. The star blows itself apart, reaching an absolute magnitude of –19 in a day or two after the initial collapse. This is more than a thousand million times the luminosity of our Sun! After reaching its peak, the brightness then rapidly falls to around absolute magnitude –16 over the next month and it then continues to fall at a fairly steady rate of about one magnitude every two months.

Type II supernovae are the final death-throes of massive stars that have come to the end of their supplies of fusible material, as already described in Section 18.4. These supernovae show some diversity in their behaviour, though their common characteristics can be described. Typical Type II events are rather less bright than Type I. They most often reach absolute magnitude values of around –17 a day or so after the core collapses and they often show a rapid fall of around a magnitude or two over the next month, remaining steady in brightness for a couple of months, then rapidly falling in brightness over the following year or so. Astronomers can also distinguish between Type I and Type II supernovae by the characteristics of their spectra.

The last supernova to be observed in our own Galaxy was seen in the constellation of Ophiuchus in 1604, before the invention of the telescope. However, astronomers have witnessed many of these spectacular stellar deaths in other star systems (galaxies – see Fig. 20.2) and recently one in a satellite system to our own – the Large Magellanic Cloud. It was discovered on 24 February 1987 by a number of independent observers. Ian Shelton, of the University of Toronto, happened to be exposing a plate of that area of the sky using a telescope at the Las Campanas Observatory in Chile. He developed the plate immediately after the exposure and found the object. Oscar Duhalde, a night assistant at the same observatory, was taking a nocturnal stroll to the coffee room when he glanced up and noticed the fourth-magnitude interloper in the Magellanic Cloud. Other independent discoveries followed, such as that by the amateur astronomer Albert Jones in New Zealand, and the supernova was soon headline news.

This supernova set astronomers something of a puzzle. The way its brightness changed with time did not conform to any of the previously documented cases. After discovery it gradually rose in brightness from magnitude 5.0 to 3.0 over a period of about 3 months. After that its brightness dwindled pretty much in the way expected for a Type II supernova. The distance of the Large Magellanic Cloud is reckoned to be 50 000 parsecs. At this distance the absolute magnitude of the object was –15.5. This was rather faint for a supernova, even one of Type II.

Astronomers were initially very surprised when investigation of pre-discovery images of the region showed that the progenitor star was a blue supergiant, and not the expected swollen red supergiant. However, this revelation does explain the supernova's behaviour.

Look again at the H–R diagram in Fig. 16.7 and find the location of the supergiants, extending from blue at the upper left to red at the upper right. Then compare with Fig. 16.8, which shows stellar radii, and you will see that blue supergiants are physically very much more compact than red supergiants. This is the clue to this supernova's odd behaviour.

The relatively small size of the progenitor (probably about twenty times the diameter of our Sun – as opposed to the usual several hundred solar diameters)

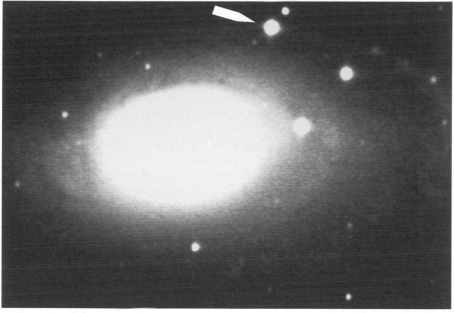

a

b

Figure 20.2 Supernova in the galaxy M81. CCD images by Martin Mobberley using his 19 inch (490 mm) Newtonian reflector. **a** 1993 April 11, showing the supernova (arrowed) at its maximum brightness (about mag 11). **b** 1994 January 17. Notice that the supernova is now much dimmer, despite this second exposure revealing fainter details in the galaxy, itself.

resulted in much of the initial heat energy released by the explosion being used to expand the star until its outer layers were thin enough to allow the pent-up energy to radiate more freely. This process took the first two or three months after the initial core collapse and this is why the supernova was dimmer than usual at the beginning and why it gradually brightened over that time.

Several neutrino detectors operating at various locations on the surface of the Earth recorded bursts of activity at exactly the same time – $7^h 35^m$ UT on the day of discovery. Theorists think that this marked the time of the core collapse that caused the outburst (bear in mind that both the light and the neutrinos took 160 000 years to cross the space separating the Large Magellanic Cloud and the Earth, so the actual event was 160 000 years earlier).

It is still a mystery why the progenitor was a blue supergiant. If proof were needed, this is ample evidence that our ideas about stellar evolution are far from being complete and accurate.

Astronomers were able to learn much about supernovae in general just because *SN 1987A*, as it is called, was so atypical.

Turning back to the Crab Nebula, its distance is reckoned at 1800 parsecs, so the actual explosion happened about 6000 years before the Chinese observed it. The nebula is expanding at the rate of about 1500 km/s and is the source of powerful emissions of radiation in radio and X-ray wavelengths. The light from the nebula is strongly polarised, indicating that it is created by electrons spiralling along magnetic field lines and losing energy by a process known as *synchrotron emission*.

A pulsar (neutron star) was found flashing in the heart of the Crab Nebula in 1968. It has a period of 0.03 second, though this is slowing down at a rate of 4×10^{-13} second per second. It is the liberation of the neutron star's rotational energy that keeps the nebula emitting in visible and other wavelengths. Without this energy source the nebula would have thinned out enough to become invisible long ago.

Another example of a supernova remnant is the Veil Nebula, which resulted from a supernova in prehistoric times.

Other supernovae exist and astronomers face the exciting prospect of watching the development of a remnant from the 1987 event. At the time of writing some nebulosity can already be seen, but so far any neutron star has evaded definite detection.

20.3 Stellar Populations

Supernovae enrich the interstellar medium with heavy elements (those of atomic mass greater than helium). In particular, all the elements heavier than iron are synthesised in the supernova blast. Stars that are subsequently born from this enriched material show the evidence in their spectra. Hence the younger a star is, the greater the proportion of heavy elements it will have. Astronomers have divided stars into two subclasses, according to their chemical compositions.

Population I stars are the youngest, with high heavy-element abundances. The Sun is an example of a population I star. *Population II* stars are the oldest, formed at a time before many supernovae had exploded their processed material into space. They contain very little in the way of heavy elements. Without this processing of the primordial hydrogen and helium the rocky planets of our Solar System could never have come into existence. Our Earth, the buildings in which we live, this book, you and I are all made of material than was once deep inside a star!

20.4 Black Holes

A star in whose core the nuclear reactions have ceased will become a white dwarf, provided it is not more than 1.4 times as massive as the Sun. If its mass is a little greater than this then it will collapse to form a neutron star. One might ask if there is any limit to the mass of a neutron star and, if so, what happens to a stellar remnant whose mass is greater than this limit?

According to our present understanding, a stellar corpse of greater than 3.2 solar masses cannot exist as a neutron star. It will be crushed under its own weight until it is no more than a minute point of matter of virtually infinite density. Furthermore, its gravitational field will then be so strong that it will close off a region of space around itself from which nothing, not even light, can escape. This is a *black hole* – certainly the most bizarre type of object in our Universe.

All the mass of the original object is concentrated at the centre of the black hole in what is called a *singularity*. For a non-rotating hole this would be an infinitely small point, of infinite density. However, the star from which it formed is very likely to rotate and the spin speed must increase as the object shrinks. A consequence of this is that the singularity will not be a point but rather a small annulus, or ring, of hyperdense matter. At a certain distance from the singularity the gravitational field strength will be such that the escape velocity equals that of light, 3×10^8 m/s. This is the *event horizon* and its radius from the singularity is known as the *Schwarzschild radius*. Newtonian mechanics can be used to calculate the Schwarzschild radius, R_s, for an object of mass M kilograms:

$$R_S = 2GM/c^2$$

where G is the universal constant of gravitation (6.7×10^{-11} Nm^2/kg^2) and c is the speed of light. Thus a 3 solar-mass black hole (the mass of the Sun = 2×10^{30} kg) would have a Schwarzschild radius of 9 km. Note that the more massive a black hole is, the greater is its radius.

Any object falling into a black hole would be lost forever. Once it crossed the event horizon it would be completely cut off from the rest of our Universe. The physics of black holes is strange indeed. Inside the event horizon the roles of space and time are reversed. It seems that it is possible to avoid destruction by the singularity if the black hole is rotating. Indeed, the black hole then appears to form a bridge between this Universe and another, or maybe to some distant location in our own Universe!

Do black holes exist? Certainly they are predicted in relativity theory, which shows no sign of being flawed. Stars of masses more than 100 times that of our Sun are known to exist, though they are very few and far between. What about their remnants – do the mass-loss processes prevent them being sufficiently massive to form black holes? If not then black holes must surely exist. Do we have any observational evidence for their existence?

One of the most powerful X-ray sources in the sky, Cygnus X-1, appears to be associated with a 20 solar-mass blue giant star. This star shows perturbations in its proper motion which can only be caused by a companion of about 10 solar-masses. However, the companion is totally invisible. If it were a normal star it would be very easily seen. It is too massive to be a neutron star, so is it a black hole? There is a growing weight of evidence to support this view. The intense X-ray flux is thought to be created by material from the 20 solar mass star falling onto an accretion disk before being swallowed up by the black hole itself. Under the powerful gravitational field of the black hole much energy would be liberated by the in-falling material, and this could explain the intense emission.

Astronomers know of other, though perhaps less incontrovertible, black hole candidates within our Galaxy. An even more exciting prospect is that some of the external systems, the galaxies beyond our own, might harbour supermassive black holes at their centres. Galaxies are the subject of the next chapter.

Questions

1   Write a short essay about *novae*, explaining their nature and the mechanism that gives rise to this phenomenon.

2   Write an essay about *supernovae*, clearly explaining the differences between *Type I* and *Type II* events, and how they are produced.

3   Explain what are *population I* and *population II* stars, and account for the difference between them.

4   Outline our ideas concerning black holes and the evidence we have for their existance.

Chapter 21

The Galaxies

A misty band of radiance girdles the sky, crossing constellations such as Cassiopeia, Cygnus and Sagittarius. We call this band the *Milky Way*. When Galileo first turned his telescopes to the skies he made a series of fundamental discoveries. One of these concerned the nature of the Milky Way. Even his tiny telescopes showed that this luminous band is composed of innumerable faint stars apparently crowded together. We now know that the Sun at the centre of our Solar System is but one member of a vast system of stars which includes all those that Galileo could see and countless more besides. This system we call the *Galaxy*.

21.1 The Shape and Size of Our Galaxy

An Englishman, Thomas Wright, was the first to explain the appearance of the Milky Way and, in the process, define the shape of our Galaxy. He thought that the stars were arranged in a flat disk and that the band of radiance arose because we were viewing it from inside. He published his theory in a book, *An Original Theory, or New Hypothesis of the Universe*, in 1750, though Wright's fellow scientists paid scant attention to this treatise.

However the noted philosopher Immanuel Kant heard of Wright's ideas and further extended them. He proposed that the Galaxy took on a disk shape because it slowly rotated and he proposed that many of the distant nebulae found by Messier might have a similar shape. In particular the oval nebula in Andromeda (M31), Kant thought, might be a distant system of stars similarly flattened by rotation. Figure 21.1 (*overleaf*) shows how M31 would appear when seen visually through powerful binoculars or a rich-field telescope. It would actually be a circular disk, but its elliptical appearance would be caused by the system being inclined at an angle to us. These ideas are surprisingly modern, though few astronomers of the time accepted them.

Sir William Herschel was a giant among observational astronomers. One of the grandest tasks he undertook was to fathom "the interior construction of the heavens", as he put it in a paper in 1784. As well as systematically examining all

Figure 21.1 Galaxy M31 in Andromeda, photographed by the author using the 13 inch (330 mm) astrographic refractor, formerly of the Royal Greenwich Observatory at their Herstmonceux site.

the types of celestial object that swept into the field of view of his telescopes, he carried out a large number of "star gauges". These were counts of the numbers of the stars visible in the field of view when his telescope was pointed in selected directions in the sky. He made the assumption that the stars are, on average, equal distances from each other. He independently came up with the same result as Thomas Wright. Jacobus Kapteyn used more refined statistical techniques in the late nineteenth century.

Herschel and Kapteyn both placed the Sun near the centre of the Galaxy. On Kapteyn's scheme the Galaxy was a flat disk 1200 light years in diameter.

Henrietta Leavitt made a crucial discovery in 1912. She found that the brightnesses and periods of Cepheid variable stars were linked in a very precise way. The brighter the star, the longer its period. Ejnar Hertzsprung built on Miss Leavitt's discovery by calibrating the brightness scale using nearby Cepheid variables. In this way observations of a Cepheid variable could be used to determine its pulsation period. This in turn fixes its absolute magnitude. Knowing the star's apparent magnitude as well as its absolute magnitude allows its distance to be determined.

Harlow Shapley capitalised on Leavitt's and Hertzsprung's work by using variable stars, actually RR Lyrae stars, to determine the distances to *globular star clusters*. These are vast spherical swarms of old stars. A typical cluster might contain from 100 000 to 1 million stars, all of which are population II. Over 100 of our Galaxy's globular star clusters are visible to us.

Knowing their distances and directions, Shapley was able to work out their three-dimensional distribution in space. He found that they formed a spherical halo around the Galaxy and were concentrated at a point in space far removed from the Sun. This point was taken to be the centre of the Galaxy. Shapley's initial results seemed to indicate that the Sun lay near the edge of the Galactic disk, which appeared to be a great deal bigger than Kapteyn had calculated. A fierce debate followed, but Shapley was right and his picture of the structure of our Galaxy was eventually accepted in the 1920s.

Obviously, a great deal more work has been done since Shapley's time to fathom the structure of our Galaxy. Astronomers now know that it is a disk 25 000 kiloparsecs (about 80 000 light years) across. It has a central bulge roughly 3 kiloparsecs in diameter and 2 kiloparsecs thick (see Fig. 21.2, *overleaf*). The main disk of our Galaxy is very thin in comparison with its diameter, being only a few hundred parsecs thick at the location of the Sun. We now know that the Sun is neither at the edge nor at the centre of the Galaxy but is about a third of the way in from the edge.

Some 200 000 million stars populate our Galaxy, but the vast majority of these are not visible to us because of the absorption of starlight by interstellar matter. This is the reason why the Galaxy appeared much smaller to the early researchers, and with the Sun at its centre. As time went by, new techniques were found to probe the make-up of our Galaxy, but instrumental in their use were the identification and study of systems external to our own.

21.2 Other Galaxies

A small telescope turned towards the nebula M51, in the constellation of Canes Venatici, will reveal a dim, fuzzy "blob", with a fainter and smaller fuzzy "blob" a few arc minutes away. Herschel, using his speculum metal reflector of 18.7 inches aperture, described the brighter "blob" as a bright condensation in a luminous disk. He also saw a faint luminous arm connecting this disk to the fainter "blob". In 1845 the third Earl of Rosse turned his 72 inch reflector to M51. The superior light-gathering power showed M51 in its true guise – the luminous disk surrounding the bright nucleus was a branching pattern of arms, swept back into a spiral arrangement. Lord Rosse found other examples of spiral nebulae and he drew them as best as he could.

Incidently, the modern telescope mirrors with their reflective coatings are much more efficient than the speculum metal mirrors of the old telescopes, and so one no longer needs such a large telescope to see the spiral form of M51. I find that I can just make it out with my $18\frac{1}{4}$ inch (46 cm) reflector when the sky is really dark and clear.

Soon photography was to supersede drawings made at the telescope. Isaac Roberts, in particular, pioneered "deep sky" photography near the end of the nineteenth century. Roberts used a 20 inch reflector to photograph the nebula M31 in Andromeda in 1887, showing its spiral pattern for the first time. Given a star-map, M31 is particularly easy to find and it is visible to the eye as a faint misty oval when the sky is dark and clear. Binoculars show it very well, revealing its bright central condensation and inclined disk, though in large telescopes it is somewhat disappointing – just a bigger and brighter fuzz. Long-exposure photography is

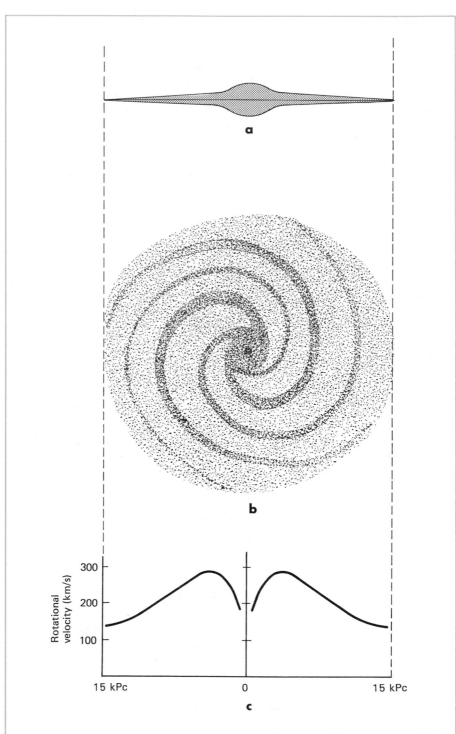

a

b

Rotational velocity (km/s)

300

200

100

15 kPc   0   15 kPc

c

needed to show it in its true guise. A modern photograph of the Canes Venatici spiral, M51, is shown in Fig. 21.3 (*overleaf*).

One of the first tasks of the 100 inch Hooker reflector, which became operational in 1918, was the photography of nebulae. Edwin Hubble found that he could categorise the large variety of distant nebulae into several distinct types. Moreover, he found that he could arrange these types into a morphological sequence. Hubble's scheme is illustrated in Fig. 21.4 (*overleaf*). It is tempting to imagine that it shows an evolutionary sequence. Hubble thought so at first, then he began to have doubts. The passage of time has proved these doubts correct.

A giant step forward was achieved in 1923 when Hubble suceeded in identifying Cepheid variable stars in M31. By observing their periods he was able to fix the distance of the nebula at three quarters of a million light years, well beyond the limits of our own Galaxy. Hence M31, and nebulae like it, were proved to be external systems, whole galaxies, like our own.

Actually, Hubble made a mistake in his calculation of M31's distance. At that time it was not realised that two classes of Cepheid variables existed – those of population I and those of population II. Population I variables, the type found by Hubble in M31, are brighter than the population II Cepheids of the same period. The distance of M31 was greater than had first been thought. We now know that it is 2.2 million light years (670 kiloparsecs) away. M31 appears big and relatively bright to us on the Earth. Many galaxies were known that appeared extremely small and dim. Astronomers realised that they must be at colossal distances from us. The Universe was turning out to be a very big place!

21.3 The Structure of Our Galaxy

Astronomers were able to learn much about the structure and composition of our own Galaxy by studying the photographs of others taken with large telescopes. However, until recent years there was no direct way of observing our own Galaxy. The interstellar matter in the plane of our Galaxy prevented astronomers from seeing more than a few kiloparsecs into it. The dark lanes cutting through the Milky Way, such as that which appears to divide it in two in the constellation of Cygnus, are the result of this interstellar absorption. The direction of the nucleus of our Galaxy is indicated by a general brightening and thickening of the Milky Way in the constellation of Sagittarius (and an intensification of the obscuring dark clouds).

Luckily, radio astronomy has come to the rescue. Radio waves can penetrate the interstellar matter. Also, neutral hydrogen gas emits at a characteristic wavelength of 21 cm. In the 1950s astronomers used radio telescopes to map the neutral hydrogen in our Galaxy. The resulting radio maps showed that it was spiral in form (see Fig. 21.2(c)). Jan Oort had already done pioneering theoretical work on the rotational dynamics of the Galaxy and his results were used in

Figure 21.2 a View of the Galaxy seen edge-on. **b** View of the Galaxy seen face-on. **c** Velocity curve of the Galaxy.

Figure 21.3 The "Whirlpool Galaxy", M51, photographed at the prime focus of the Isaac Newton Telescope, formerly at Herstmonceux. RGO photograph.

constructing the map. The Sun and its attendant Solar System move round the centre of the Galaxy in a nearly circular orbit, with a velocity of about 250 km/s. It takes roughly 200 million years to do one complete circuit.

The stars closer to the centre of the Galaxy move with greater velocities and those further out move more slowly. Sir William Herschel, in a paper to the Royal Society in 1783, determined the proper motion of the Sun with respect to the nearby stars. The principle of his method is simple. Imagine driving along a well-lit road at night. If you glance behind you see the street lamps apparently closing together. If you glance sideways you see the lamps rushing past in a

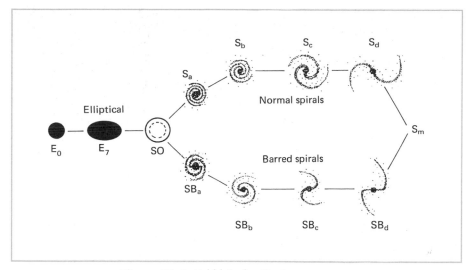

Figure 21.4 Hubble's classification sequence.

direction opposite to that which you are moving. Looking ahead you see the oncoming lamps apparently opening out from a point ahead of you. Similarly, we see the motion of the Sun reflected in shifts in positions of the nearby stars. Of course, we have already seen that the stars undertake a variety of different sorts of motions but, in principle at least, we can sort out the shifts due to the motion of our Sun.

Herschel, using data published by Maskelyne, the Astronomer Royal for England at the time, found that the stars appear to be rushing away from a point in the sky within the constellation of Hercules – the *solar apex*. This, he correctly reasoned, is the direction of travel of the Sun and Solar System.

Our Galaxy is a typical example of a spiral galaxy. It has three distinct parts – the disk, the central bulge, and the halo. The disk is very thin and flat, as already noted. It contains mainly young (population I) stars, together with a great deal of gas and dust from which new stars are being born. The interstellar matter and the youngest stars are concentrated along the spiral arms. Colour photographs of spiral galaxies show their arms to be predominantly blue, because of the highly luminous hot blue stars they contain. The spiral arms are also speckled with red HII regions. Ours would appear the same if we could see it from outside.

By contrast, the central bulge of our Galaxy is chiefly composed of old (population II) red giant stars with little in the way of gas and dust. The mix of stars in the bulge give it a predominantly orange colour. The distribution of mass with radius from the centre of the Galaxy is such that the stars do not move in Keplerian orbits. Instead they follow a velocity curve, as illustrated in Fig. 21.2(c). Stars closer than about 8 kiloparsecs from the centre actually move more slowly the closer they are to the central nucleus. The stars are also more densely packed near the centre. At a distance of 100 parsecs from the nucleus the average star density is about 100 solar masses per cubic parsec. This is fifty times the star density in our location. At 10 parsecs from the centre this value has increased to

7000 solar masses per cubic parsec. The stars here are only about one-tenth of a parsec apart. Imagine the sight of the night sky from a hypothetical planet orbiting a star near the Galactic nucleus. The heavens would be ablaze with brilliant pinkish stars, turning night almost into day!

As mentioned earlier, the halo is composed of globular star clusters. In these objects the stars are packed with densities similar to that in the nuclear regions of the Galaxy. Their mutual gravitation stops a cluster from falling apart, though some of the outer stars are probably lost as it passes through the disk of the Galaxy twice every orbit.

From our viewpoint, the grandest globular star clusters in the sky are ω Centauri and 47 Tucanae, though they lie too far south to be seen from northern latitudes (their declinations are $-47°$ and $-72°$, respectively). Third best is M13 (see Fig. 21.5). M13 is visible from the northern hemisphere, in the constellation of Hercules. It can just be seen with the naked eye on a really clear night and binoculars show it easily. A telescope of around 3 inches aperture is able to resolve some of the stars in its outer portions. I have often succumbed to the temptation of turning my own telescope, or a larger instrument, towards M13. It is a truly stunning spectacle when seen through a large telescope, with virtually the entire field of view dusted with stars, and becoming a glorious blaze of light at its centre.

Figure 21.5 The globular star cluster M13, photographed using the Isaac Newton Telescope, formerly at Herstmonceux. RGO photograph.

21.4 Galactic Morphology

Galaxies come in a variety of different shapes and sizes. I have already mentioned Hubble's classification sequence. On this scheme types E0 to E7 are elliptical galaxies, with E0 being spherical and E7 the most elongated. They are dominated by old red giant stars and so are yellowish in colour. They also resemble the nucleus of our Galaxy in that they are relatively dust free.

Types Sa, Sb, Sc and Sd are spiral galaxies with increasingly open spiral arms and less prominent central bulges, progressing from Sa to Sd. Figure 21.6 shows NGC 7331 an Sb galaxy in Pegasus, while Fig. 21.7 (*overleaf*) shows M66 – also an Sb galaxy, though with slightly more open spiral arms. Types SBa, SBb, SBc and SBd correspond to to types Sa, Sb, Sc and Sd, but have the additional feature of a straight bar of interstellar matter and stars passing through the nucleus of the galaxy. They are known as *barred spiral galaxies*, as opposed to the other type – *normal spiral galaxies*. Spiral arms trail off from the ends of the bar. Figure 21.8 (*overleaf*) shows NGC 7479, a fine example of an SBb spiral.

As already mentioned, the light from the disks of spiral galaxies is dominated by blue population I stars, while most of the light from the nuclear regions originates from the older population II stars, and is orange–yellow as a result. Spiral galaxies tend to have strong concentrations of gas and dust in the galactic disks. Radial velocity measures have shown that they rotate with their spiral arms trailing, confirming the visual impression.

Type S0 are called *lenticular galaxies*. They are, as their name would imply, lens-shaped. They have large central bulges, but the disk is only weakly featured. Their stellar make-up closely resembles that of elliptical galaxies. No spiral arms

Figure 21.6 The galaxy NGC 7331 in Pegasus. CCD image by Martin Mobberley, using his 19 inch (490 mm) Newtonian reflector.

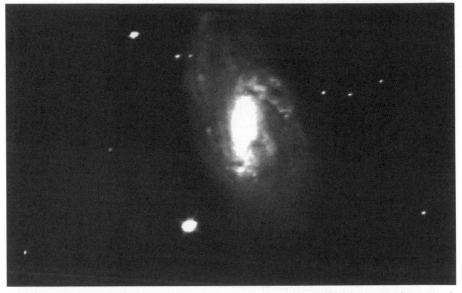

Figure 21.7 The galaxy M66 in Leo. CCD image by Martin Mobberley, using his 19 inch (490 mm) Newtonian reflector.

Figure 21.8 The galaxy NGC 7479 in Pegasus. CCD image by Martin Mobberley, using his 19 inch (490 mm) Newtonian reflector.

are identifiable in the disk. *Irregular galaxies* form a sort of miscellaneous class. Most examples have wildly distorted and chaotic shapes and they are typically rather small. The Magellanic Clouds, two mini-galaxies that orbit our own, are both irregular.

The masses of galaxies range from a few hundred thousand times the mass of our Sun, for the so-called *dwarf galaxies*, to 10 million million solar masses for the giant ellipticals. The spiral galaxies tend to have smaller masses than the ellipticals. Our own Galaxy is an Sb spiral with a mass 200 thousand million times that of our Sun, a little larger than the average for this type of galaxy. A census of galaxy types reveals that about 60% of the galaxies are elliptical, with 30% of them spiral and 10% irregular.

One or two per cent of the galaxies do not fall neatly into the foregoing classification scheme. They are termed *peculiar galaxies*. They arise because of the motions of galaxies in clusters. Every once in a while one galaxy passes close enough to tidally distort another. The galaxy M51 (NGC 5194), pictured in Fig. 21.3, has been distorted by the close passage of NGC 5195. This second galaxy is now connected to the large spiral by a linking arm of gas and stars, drawn off as a result of the interaction. A few galaxies undergo direct collisions, resulting in fantastic deformations and sprays of gas and stars shot into the intergalactic void.

Globular star clusters are identifiable on the photographs of many nearby galaxies, and astronomers think that they are the natural attendants to most galaxies.

21.5 The Formation and Evolution of Galaxies

Astronomers have long sought to understand how the galaxies have formed and evolved. At present their ideas remain very theoretical. However, recent advances have been made using powerful computers to enact simulations of the dynamical development of large amounts of matter.

If we start from a fairly uniform distribution of gas filling space, then dynamical instabiities will tend to cause vast clouds, of roughly a million million solarmasses, to form. This gas will gradually contract and form a central nucleus. The condensation will continue and stars will start to form over a time-scale of about 100 million years. If nearly all the available gas is used up in forming stars, the star formation will virtually cease and the result will be a large spherical or elliptical mass of stars – an elliptical galaxy. The galaxy will be mainly composed of population II stars, and this model predicts that we will see little evidence of gas and dust in them today. This agrees with what we do observe.

However, if the rate of star formation proceeds more slowly there will be sufficient time for some of the matter to fall into the plane of rotation of the cloud and so form a disk there. The resulting galaxy will consist of a central bulge of existing stars, a disk of matter and newly forming stars, and a halo of stars (thinly spread) and globular star clusters. This is just what we do see in lenticular and spiral galaxies. Astronomers estimate that our own Galaxy is about 10 000 million years old, roughly twice the age of our Sun.

Of course, the initial conditions in the pre-galactic cloud will have lot to do with the way a galaxy evolves. Gas turbulence and temperature, as well as the mass of the initial cloud, must play an important role. Some clouds will give rise to massive elliptical galaxies. At the opposite extreme, other clouds will form into dwarf irregular galaxies.

A vexing problem which astronomers have had to face is how the bars and spiral arms possessed by many galaxies are formed and maintained. The first point is that the stars are **not** arranged in fixed patterns in the spiral arms. A galaxy does not revolve as a solid object. As already mentioned, the individual stars move with velocities that depend on where they are in the galaxy. As a result, the stars in a galaxy's disk pass in and out of the spiral arms as they move along their orbital paths. At present we lie inside one of the spiral arms of our own Galaxy, though our Sun has spent much time in the emptier regions between arms in the past. It will do so again in the future.

Computer simulations have shown that gravitational instabilities will tend to develop in a galaxy's disk and these lead to compressional waves, called *density waves*, in the matter in the disk. These density waves tend to cause bunching of the stars and gas. In turn this promotes new star formation, which is why we see the hottest and youngest stars in the spiral arms. As the simulations are run they show that the spiral arms disperse after a few rotations of the galaxy and a new density wave in the form of a central bar develops. This is a notable success in achieving an explanation of these peculiar features, but the theory is not without its problems. For instance, why do many spiral galaxies apparently never develop bars and, in these cases, what maintains the spiral patterns?

The computer simulations show that a bar might not develop if the galactic halo is very massive. So far astronomers have found little direct evidence of this extra mass (though there are now indications of much unseen matter permeating space – but more of this in the next chapter). However, they do have a good mechanism for explaining the longevity of the spiral arms. The first and most massive of the disk stars to form will be the first to explode in supernova outbursts. These outbursts will cause shock waves to sweep through the interstellar medium, compressing it and triggering the formation of new stars. The most massive of these will then erupt as supernovae, creating new shock waves which lead to further stars and, eventually, more shock waves.

The shock front from each successive supernova blast will thin out as it moves away from the explosion centre, eventually becoming ineffectual a few hundred parsecs away from where it began. The triggering of new star formation continues as a chain reaction, like a sort of interstellar relay race, and the differential rotations in the galactic disk result in the star propagation spreading out along spiral paths. Again, computer simulations of this mechanism in action provide a good match to the actual appearances of spiral galaxies.

A galaxy starts life by being composed of only hydrogen and helium. As it ages the amounts of heavy elements increase due to processing, known as *nucleosynthesis*, by the stars. Also the light from the galaxy gradually reddens as the number of red giant stars increase.

21.6 Groups of Galaxies

Most galaxies do not exist in isolation. They tend to be arranged into small *groups* or larger *clusters*. The groups tend to be a couple of million parsecs across and contain a few to a few dozen galaxies. Our own Galaxy is a member of a system we call the *Local Group*. It comprises of over thirty individual galaxies, most of which are dwarfs, with less than a hundredth the mass of our Galaxy.

The "heavyweight" members of the Local Group are our own Galaxy, the Andromeda Spiral M31, and M33 in Triangulum. M31 has a mass of 3×10^{11} times that of the Sun, about 50% greater than that of our own Galaxy. It is the largest and most massive member of the Local Group. The disk of this galaxy spans 50 kiloparsecs. It is an Sb type spiral and has a number of small satellite galaxies, which orbit close to it. The most prominent of these are the two elliptical galaxies M32 and NGC 205 (this latter has no Messier catalogue number), both just about visible in Fig. 21.1. The spiral arms of M31 are not easy to appreciate, because the Galaxy is strongly inclined to us. Its disk makes an angle of about 13° to our line of sight.

By contrast, we see M33 almost "face-on". It is an Sc type spiral of mass 1×10^{10} times that of the Sun, with a diameter of 8 kiloparsecs. It, too, has dwarf companions. The remainder of the Local Group consists of small irregular and spheroidal galaxies, most of which are clustered around the "big three". Other groups of galaxies lie a few megaparsecs from each other.

Dotted among the groups of galaxies are the much larger and more massive clusters. The closest rich cluster to us is that in Virgo, about 16 megaparsecs away. It is a vast system of several thousand galaxies, over 3 megaparsecs in diameter. In fact, the Virgo cluster forms the nucleus of a collection of galactic groups, about 18 megaparsecs across, which we call the *Local Supercluster*. Our Local Group is a member of this supercluster, situated near its periphery.

It has now become apparent to astronomers that almost all galaxies are members of groups or clusters (those few that are not are known as *field galaxies*) and virtually all groups and clusters are members of superclusters. Moreover, great volumes of empty space exist between the superclusters. The results of the most recent research seem to indicate that the galaxies and clusters are spread through space in a sort of cellular structure. If you imagine a pile of soap bubbles, the air inside the bubbles might represent the empty volumes of space, while the walls of the bubbles would correspond to the regions where the galaxies are concentrated. The points of intersection between bubbles would then correspond to the superclusters. The size of a "cosmic bubble" is of the order of a hundred megaparsecs. On scales larger than this we think that the Universe is fairly homogeneous.

21.7 The Expansion of the Universe

An early pioneer of spectrography applied to the galaxies was Vesto M. Slipher, of the Lowell observatory. He had to use very long exposures with specially constructed equipment, since the galaxies are all rather dim objects. To collect the light Slipher used the 24 inch refractor which Lowell had erected for his studies of Mars. He often had to continue exposures on one particular object for several nights in order to build up the faint spectrum on a photographic plate. Even then, the spectra were only of low dispersion. However, Slipher succeeded in making a remarkable discovery. He found that the light from these distant objects was measurably Doppler-shifted to the red end of the spectrum. In other words, the spectra indicated that the galaxies were receding from us at high velocities. By 1925 Slipher had managed to obtain the radial velocities of forty nearby galaxies. He found recessional velocities up to 1800 km/s!

Edwin Hubble wondered whether the Universe itself was expanding, carrying the galaxies along with it. Between 1928 and 1936 Hubble and an assistant, Milton Humason, worked with the Hooker reflector, obtaining the spectra of galaxies to investigate this theory. Hubble knew the distances of the nearby galaxies from measurements of the Cepheid variable stars, but he had to use other techniques for those more distant. In particular he used the apparent brightnesses of galaxies as indicators of their distance.

By 1929 he was in a position to announce that the recessional velocity of a galaxy is proportional to its distance from us. This relationship is fundamental to our understanding of the Universe and has become known as *Hubble's law*. It can be expressed mathematically as:

$$v = Hd$$

where v is the velocity of the galaxy, usually expressed in km/s, and d is the corresponding distance of the galaxy in megaparsecs. H is then the constant of proportionality, known as *Hubble's constant*. Astronomers currently believe that the value of H lies between 50 and 100 km/s per megaparsec. The large degree of uncertainty in the value of H arises because we still have no very accurate way of independently fixing the distances of any but the nearest galaxies.

It is quite easy to understand how Hubble's law proves that the Universe is expanding. Imagine blowing up a balloon that has a polka-dot pattern on its surface. As the balloon is inflated, so the dots get further apart. If you imagine being on one dot and viewing the others, you can see all the dots are moving away from you. Also, the dots furthest away appear to move the fastest. Exchange the dots for the galaxies and you get the idea of what is happening to the Universe. The galaxies are not all running away from us because our Galaxy is at the centre of the Universe, or is in some other way unique. All the galaxies share in the expansion. From any one galaxy all the others would appear to be receding (neglecting individual motions within groups of galaxies – actually several members of our Local Group are approaching us). The greater the distance of the galaxy, the greater its *recessional velocity*. Figure 21.9 shows the general idea.

Of course, if Hubble's law is taken to hold true (and recessional velocities are taken to be the only physical cause of redshift) then redshift measures can be used to give the distances of the most remote galaxies. By 1936 Humason had recorded the spectrum of a galaxy that was moving away from us at 40 000 km/s. If we adopt a value for H of 80 km/s per megaparsec, then this distance corresponds to 500 megaparsecs (1600 million light years). This was the limit of the 100 inch telescope with the photographic plates and equipment in use at the time. Today, astronomers can measure the redshifts of galaxies very much further away than this. How far – and how big is the Universe, anyway? These are questions which will be discussed in the next chapter.

Figure 21.9 The expansion of the Universe. **a** and **b** schematically represent the sizes of the Universe at an early and later epoch, respectively.

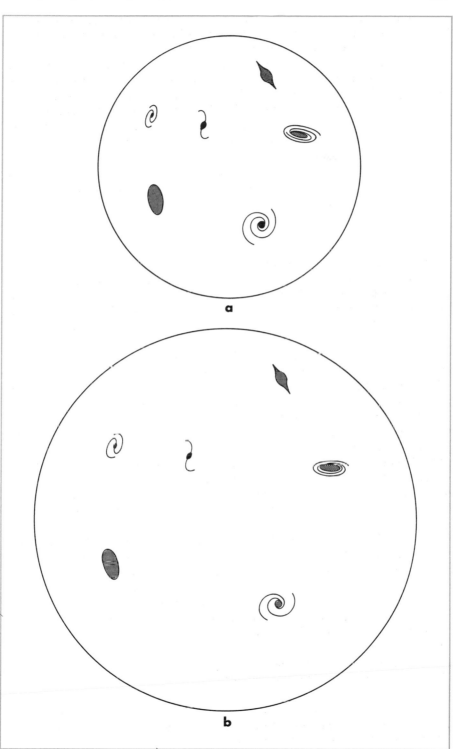

a

b

Questions

1   Explain the cause of the band of radiance in the night sky known as the *Milky Way*.

2   Give an account of how astronomers arrived at a knowledge of the shape and size of our Galaxy. State the modern figures for the dimensions of our Galaxy and the number of stars it contains.

3   Draw the *Hubble sequence* of galaxy types and give the main physical characteristics of each type of galaxy.

4   Explain how radio astronomy has been used in probing the structure of our Galaxy.

5   What type of galaxy do we live in? How are the different stellar populations distributed within it? Describe how the Galaxy would appear as seen from outside – giving particular details of its colour.

6   Describe the nature of *globular star clusters*. How are our Galaxy's globular star clusters distributed?

7   Describe the modern theory of how a galaxy forms. Explain how the various morphological types (such as ellipticals and spirals) arise on this scheme.

8   Describe the current ideas of how the spiral arms of galaxies are formed and maintained.

9   Give an account of the large-scale arrangements of galaxies in space.

10 Write an essay about the discovery of galactic *redshift* and the expansion of the Universe.

Chapter 22

Quasars, Active Galaxies and Cosmology

THE 1960s was a period of major upheaval in astronomy. Particularly significant was the discovery of immensely powerful objects well beyond the most distant of the detected galaxies – the quasars.

22.1 Quasars

By 1960 radio astronomers had found and catalogued several hundred radio sources in the heavens. Optical astronomers were able to show that some of these were identifiable with optical objects, usually galaxies. They became known as *radio galaxies*. However, the majority of these objects did not appear to have any optical counterparts.

In 1960 two optical astronomers, Allan Sandage and Thomas Matthews, found a sixteenth-magnitude, bluish, star-like object at the position of the forty-eighth radio source in the Third Cambridge catalogue. 3C 48 proved to have a peculiar spectrum. It showed several broad emission lines that the astronomers could not identify with known elements and its ultraviolet intensity was very much greater than any normal star of the same brightness. Further observations provided another surprise. It was found to fluctuate in brightness. It changed its brilliance by more than a third over the course of a year. Astronomers thought that 3C 48 was a peculiar, and unique, radio-emitting star situated within the confines of our Galaxy.

However, in 1963 another strong radio source, 3C 273, was identified with a thirteenth-magnitude bluish star, showing the same sort of peculiar spectrum. Maarten Schmidt, working at Mount Palomar, realised that the peculiar spectrum of 3C 273 could be explained as the characteristic lines of hydrogen redshifted by 15.8%. In other words all the wavelengths of the spectral lines were increased by this amount. Astronomers would usually refer to this redshift value as 0.158. He

309

took a look at the spectrum of 3C 48 and realised that its spectrum was redshifted by an even greater amount – 0.367. Remembering the formula that relates radial velocity and the fractional shift in wavelength (given in Section 15.7), it should be clear that such large redshift values correspond to recessional velocities that are a significant fraction of the speed of light. Clearly the objects could no longer be regarded as residing in the Galaxy. Moreover, if they obey Hubble's law then they must be immensely remote.

The standard formula given for the radial velocity as determined for the wavelength shift ceases to be accurate when the emitting object is moving at a speed close to that of light. A more refined formula has to be used. Even allowing for relativistic changes in mass, length and time (those changes predicted as a result of Einstein's relativity theory), which affect the perceived wavelength of radiation from a fast moving source, is not enough. We also have to allow for the change in the size of the Universe since the time when the light was emitted. These matters are studied in detail later in this chapter. Suffice it to say here that it is possible to have redshifts of greater than 100% (impossible according to the standard formula – an object could only obtain a redshift of 100% if it was receding from us at the speed of light, and nothing can move faster than that!)

Coming back to 3C 273 and 3C 48, the distances of these objects turn out to be over 600 megaparsecs (2000 million light years) and over 1400 megaparsecs (nearly 5000 million light years) respectively. For these objects to appear as bright as they do despite their immense distances, they must be exceedingly luminous. Over the years astronomers have discovered several hundred of these "quasi-stellar radio sources", or *quasars*. Most of them fluctuate in intensity. Investigations by Sandage showed that only a small fraction of the known quasars are strong radio emitters. However, all have very large redshifts and they are all exceedingly energetic objects. A typical quasar has a power output 100 times as great as a large normal galaxy!

One striking point about the quasars is that they are all very remote objects. This means that we are seeing them thousands of millions of years ago. Indeed, the numbers of quasars observed of greater redshift values increases much faster than one would expect for a uniform distribution. In other words, quasars seem to be chiefly associated with an epoch long ago when the Universe was much younger. They are certainly the most violent events we see in the entire Universe.

The fact that many of the quasars show rapid changes in optical and radio brightness indicates that the regions within the objects that produce the power output are very small, certainly less than a parsec in diameter. What mechanisms could release such excessive amounts of energy in so small a volume? Astronomers were puzzled. A possible clue comes from the discovery of many galaxies that show a "watered-down" version of a quasar's behaviour. These objects are called *active galaxies*.

22.2 Active Galaxies

Before the 1960s radio astronomy was hampered by the poor angular resolution that resulted from the long wavelengths of the radio waves. Even the newly completed "Mark 1" radio telescope at Jodrell Bank, with its 76 m diameter dish, had a radio resolution inferior to the eye working at visible wavelengths. However, the advent of the interferometer (see Chapter 4) was to change all that. For the first time radio astronomers could examine radio sources in the same sort of detail to which optical astronomers had become accustomed.

Many of the radio galaxies were found to have a common radio structure. As well as a central emitting region whose position coincided with the visible galaxy, two massive "lobes" of radio emission extend in opposite directions from the central object. Figure 22.1 shows the sort of radio map obtained from one of these objects. These lobes are the largest single structures in the observed Universe. The present record-holder belongs to the radio source 3C 236. Its lobes extend across 5 megaparsecs (over 15 million light years)!

These lobes appear to be formed by millions of solar masses of plasma ejected from the galaxy at speeds that are an appreciable fraction of the speed of light. The fast-moving plasma generates a magnetic field, and the charged particles within it spiral along the magnetic field lines releasing large amounts of energy as synchrotron emission. These *extended radio sources* tend to be unvarying in their output, but the *compact radio sources* fluctuate irregularly over relatively short periods of time. They are the objects at the centre of the extended radio sources, but many compact sources exist without any apparent lobes.

Some radio galaxies exhibit jets of material extending from their nuclei. These jets are seen in visible light as well as by their radio emissions. The brightest galaxy in the Virgo cluster, M87, shows a jet of this type. The galaxy was identified as a strong source of radio emissions in 1949, and it was given the designation Virgo A. These jets are very much smaller than the radio lobes possessed by many radio galaxies, having lengths in the kiloparsec range.

In 1943 Carl Seyfert had optically identified a number of galaxies with very bright nuclei. The spectrum of a normal galactic nucleus shows a continuum of

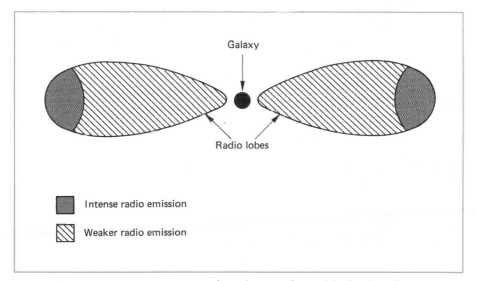

Figure 22.1 A representation of a radio map of a two-lobed radio galaxy.

black body type crossed by dark absorption lines, caused by the amassed light of large numbers of stars. However, the nuclei of these *Seyfert galaxies* displayed curious spectra dominated by broad emission lines superimposed on a continuum produced by synchrotron emission. In the following years radio astronomers found that Seyfert galaxies were powerful radio sources.

In 1968 Maarten Schmidt identified one particular compact radio source with the distant elliptical galaxy BL Lacertae. It seemed to fluctuate even more wildly than Seyfert galaxies and its optical spectrum proved to be very odd. Only a continuum emission could be seen, with no absorption or emission features visible at all! Other examples of these *BL Lac* objects were found and we now know of over a hundred. BL Lac and Seyfert galaxies are all characterised by very bright nuclei. Many of these objects seem to be as powerful as the weakest quasars. Another link is that many quasars have recently been shown to be surrounded by elliptical masses of stars.

In other words, quasars are the most extreme examples of galaxies with active nuclei. In the case of the quasars the nucleus outshines the rest of the galaxy. Astronomers might at last understand what a quasar is, but it certainly cannot be said that they yet fully understand the mechanism that produces such colossal outpourings of energy, although one idea is now gaining popularity.

The normal nuclear processes occurring in stars are rather inefficient in that only a tiny fraction of the mass of a star is converted into energy. A large normal galaxy would shine because of the conversion of something like 0.005 solar masses per year into energy (the total power output of the galaxy would be of the order of 1×10^{39} watts). In order to explain the output of a typical Seyfert galaxy one would have to envisage the conversion of 0.05 solar masses per year. Quasars would require at least ten times this mass each year. Moreover, the energy-producing regions of these objects are less than a parsec across. So how do these objects produce their energy? Certainly normal nuclear reactions are totally inadequate.

Many researchers tried to explain the power sources of these objects in a large variety of different ways, none of which are really satisfactory. Recent theories seem more promising. Currently many astronomers think that the activity in active galaxies might be caused by the presence of a supermassive black hole in their nuclei. A black hole of a million solar masses would certainly gobble up stars and interstellar matter at a rate sufficient to release the amounts of energy observed. In the process the black hole would increase its radius, drawing in yet more matter.

Dynamical studies of the motions of stars near galactic nuclei show that very large concentrations of matter are indeed present. However, even if black holes are responsible for the outputs of quasars and the other less powerful active galaxies, we still have some questions that need answering. For instance, why are only 10% of the visible quasars strong emitters of radio radiation? What produces the bland spectra shown by BL Lac objects? Why is it that not all galaxies show quasar-like behaviour?

On the last point, the centre of many apparently normal galaxies, including our own, do emit significant amounts of radio and other electromagnetic radiations. Though these emissions are not in the same league as those from the active galaxies, some astronomers think that massive black holes might be common in the nuclei of the larger galaxies.

A general theory is beginning to emerge which might explain the diverse phenomena we have discussed simply as different aspects of the same thing. The key to it all is the presence of a black hole as the central power house in active galaxies.

22.3 Black Holes, Accretion Disks, Jets and Cosmic Chameleons

A supermassive black hole in the centre of a galaxy will have a doughnut-shaped accretion disk of in-falling material in orbit around it. The material close to the event horizon of the black hole will be at a temperature of perhaps hundreds of thousands of degrees. That in the outer, thicker, part of the doughnut will be much cooler – perhaps having a temperature of a few hundred degrees. The main part of the doughnut will be a couple of light years across and will be very opaque.

The super-hot material orbiting in a thin sheet close to the black hole will be in the form of a plasma. The circulating plasma will generate a powerful magnetic field (electric charges in motion naturally give rise to a magnetic field) in the manner shown in Fig. 22.2 (*overleaf*). As the diagram illustrates, in the directions perpendicular to the plane of the plasma disk the magnetic field lines will extend in straight lines out into space. These form "fast-lane highways" for escaping electrified particles from the high-energy regions around the black hole. In this way, astronomers think, the once mysterious jets that emanate from active galaxies are a natural consequence of an accretion disk around a supermassive black hole. Such accretion disks and their associated jets have now been observed by the Hubble Space Telescope.

Of course, the material streaming (and spiralling) along the jets would plough into the surrounding interstellar medium and one might expect radio signatures of the interactions that would result. We see these, too: they are the radio lobes. Observations over the last few years have shown that the optical jets and the inner parts of the radio lobes are aligned, proving their common origin. The outermost parts of the radio lobes are often distorted and displaced, but this is to be expected from relative motions in the interstellar medium.

So, if we see a galaxy with a highly active nucleus more or less edge-on to us then it will appear like a typically radio-noisy quasar – less active and we might call it a Seyfert galaxy.

The jets emitted by an active galaxy's nucleus would be expected to produce very bland spectra, in fact, exactly the sort that are typical of a BL Lac object. It seems likely that BL Lac objects are the same sort of active galaxy as just described. They only appear different because of the different orientations which we see them. In effect we are looking straight down one of the active galaxy's jets and the result we call a BL Lac object.

If we see the active galaxy at an intermediate angle then we may expect to observe it as having intermediate properties between the two extreme cases. Examples of these intermediates abound. In effect, active galaxies are cosmic chameleons!

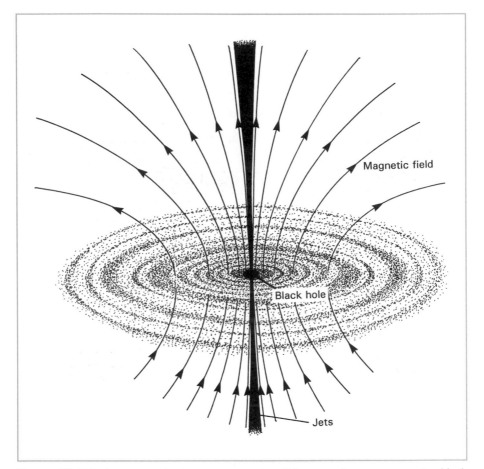

Figure 22.2 Jet-forming mechanism in an accretion disk surrounding a supermassive black hole. The magnetic field is most concentrated above and below the black hole, in a plane perpendicular to the disk, and the field lines extend straight out into space at these points. Particles can escape, and spiral along the field lines most easily at these locations, so forming the jets – which show strong synchrotron emission as a result.

Figure 22.3 illustrates what appears to be a very likely theory for unifying the apparently diverse behaviour of some of the most remote and most powerful objects in the Universe.

Figure 22.3 The different guises of an active galactic nucleus. AGN 1 would be observed from Earth as a classical two-lobed radio source. If the emission was particularly intense we would classify it as a radio-bright quasar. We are looking straight down one of the jets of AGN 2. We would see an optically bright and rapidly varying source, whose spectrum would be dominated by synchrotron emission. We would classify AGN 2 as a BL Lac object. AGNs at different angles of presentation to us would produce the wide range of intermediate cases we do see in practice. The other variable is the activity of the nucleus, which would depend on the mass of the supermassive black hole and the rate at which its consumes new material from the host galaxy.

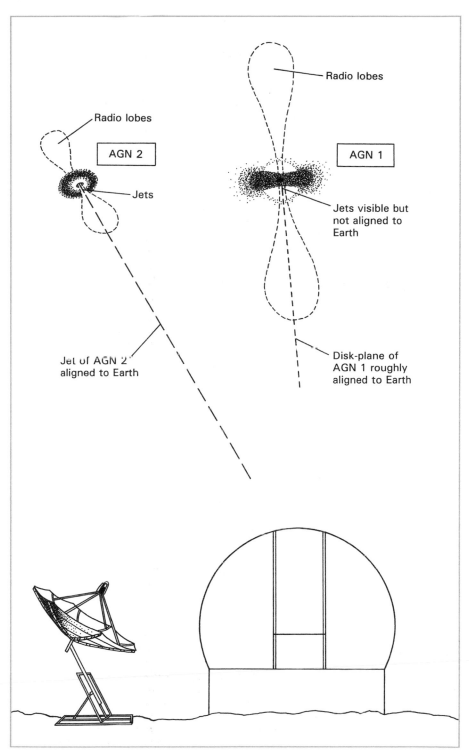

Radio lobes

AGN 2

Radio lobes

AGN 1

Jets

Jets visible but
not aligned to
Earth

Jet of AGN 2
aligned to Earth

Disk-plane of
AGN 1 roughly
aligned to Earth

22.4 Cosmology

The term "cosmos" comes from the Greek word for "order" and "everything". Nowadays, we call the study of the structure and evolution of the Universe as a whole *cosmology*.

Shortly after Albert Einstein published his *General Theory of Relativity*, many theoreticians attempted to obtain cosmological solutions to his field equations. In 1917 the mathematician Willem de Sitter announced one solution which predicted that the whole Universe ought to be continually expanding. His theory was a little primitive in that the solution was only strictly true if the Universe was empty of matter. However, it was a start and other workers improved on the theory. In particular, the Russian Alexandr Friedmann and Belgian Abbé Lemaître produced more workable versions of the theory. Hubble eventually provided the observational proof that the Universe was indeed expanding.

In the 1940s George Gamow, then working in America, took the theorists' ideas and moulded them into a scenario for the birth and subsequent evolution of the Universe. The basis for his idea is straightforward. If the Universe is expanding, then at some time in the past it must have been very small. He proposed that the Universe began with a small volume of concentrated radiation and elementary particles, the "primeval atom", which exploded to give birth to the Universe we know. Gamow's idea became known as *the Big Bang theory*. Gamow and other scientists have tried to work out the consequences of this theory and explain how the Universe came into being and developed into its present form. This work is still continuing.

However, some scientists did not like the idea of a Universe that had a definite birth. Questions such as "What happened before the Big Bang?" spring to mind. In 1948 Fred Hoyle, Hermann Bondi and Thomas Gold offered the scientific world *the Steady-State theory*. On their scheme the Universe has always existed and always will. As the Universe expands, so matter is continually created in the space between the expanding galaxies. They predicted that the Universe of long ago should appear, on the large scale, the same as it does today.

The Steady-State theory became weakened when radio astronomers discovered that radio galaxies, and particularly the quasars, were much more common in the remote past than today. Clearly the early Universe was **not** the same as it is today. The supporters of each of the theories engaged in frantic arguments in one of the fiercest debates science has ever known. However, the Steady-State theory was effectively disproved in 1965 as the result of a serendipitous discovery by two radio astronomers, Arno Penzias and Robert Wilson. They were interested in trying to pick up faint radio emissions from the outer reaches of our Galaxy. In order to do this they needed a particularly sensitive radio antenna. They used a steerable horn-shaped receiver, situated at Holmdel in New Jersey. During their research they detected a faint "hiss" of radio radiation, of wavelength 7 cm, that appeared to be coming equally from all directions in the sky.

At first Penzias and Wilson suspected the equipment, but further investigation eliminated this possibility. Various scientists had predicted that such a background would result from the afterglow left by the Big Bang. Gamow himself had predicted this in 1948. Penzias and Wilson contacted a research group at Princeton University to see if they could help. In fact this group, led by Robert Dicke, had just set up an antenna to look for this very radiation! Various workers in the years that followed have measured the microwave background radiation at various wavelengths. It proves to have a distribution of intensity with wavelength that corresponds to a black body at a temperature of 2.7 K.

The supporters of the Steady-State theory were unable to sustain their theory against this devastating evidence. In time they abandoned it. It was a case of "Big Bang rules 2.7 K".

From the present rate of expansion of the Universe, it is reckoned that it began something like 15 000 million years ago. At 1×10^{-43} second into the age of the Universe all its matter existed as a ball of incredibly hot and dense gas, little more than a singularity, with a temperature of 1×10^{32} K. In fact, since relativity theory says that time, space and matter are mutually linked, there is no such thing as time or the Universe before the Big Bang. The entire Universe was this "primeval atom".

The behaviour of radiation and elementary particles dominated the early Universe because it was so hot. At that time the Universe was an opaque void. As the Universe expanded it cooled until, about a million years after the Big Bang, it had an average temperature of about 10 000 K and a density of roughly 1×10^{-18} kg/m^3. The behaviour of matter then became dominant and the Universe was no longer opaque to the passage of radiation. It is from this phase that we see the microwave background. Effectively the radiation from this era has become red-shifted by the expansion of the Universe. The amount of redshift, z, is described by the equation:

$$z = \left(\frac{S(t_2)}{S(t_1)} \right) - 1$$

where $S(t_1)$ and $S(t_2)$ are the scale-sizes of the Universe at the time of radiation leaving an object and at present, respectively. The most distant object discovered to date is a quasar with a redshift of 4 ($z = 4$). Thus light left this object when the Universe was one-fifth of its present size, roughly 12 000 million years ago. Similarly, the 10 000 K radiation at the time when the Universe became transparent has been redshifted to the value of 2.7 K by its subsequent expansion.

Results from the *COsmic Background Explorer* satellite, or *COBE*, in the early 1990s have supported the Big Bang idea and has added further details. Certainly the black body nature of the 2.7 K microwave background radiation has been confirmed to a very high degree of accuracy. Also revealed is a very slight "lumpiness" or variations in the temperature of the background. These "lumps" correspond to structures several hundred million light years across in the Universe at around the time of the formation of the first galaxies. It appears that the Universe at that time was highly turbulent, and this leads scientists to think that galaxies interacted with each other very much more frequently then. Perhaps this accounts for the greater prevalence of quasars in the early Universe.

The correct general scenario for the evolution of systems of galaxies may be that individual spirals were the first to form and that the other galactic morphologies resulted from interactions, which also built up the clusters of galaxies and ultimately the superclusters. This is a "bottom-up" scenario, as opposed to the previous "top-down" idea. We do not know which is correct. Much more work needs to be done before we can be sure on this matter.

Gravitational instabilities developed in this expanding cloud of matter and the galaxies and clusters of galaxies eventually formed from local condensations. The result is the Universe of at least a million million galaxies all rushing away from each other that we see today.

22.5 Where Will It All End?

There are two possibilities for the eventual fate of our Universe. One is that it will go on expanding forever. If this is the case then eventually the last star in the last galaxy will cease to shine and the Universe will thin out into a "nothingness" of dead matter and heat radiation, becoming ever more diluted.

On the other hand the expansion of the Universe might one day cease (on the principle of "what goes up must come down") and the Universe of matter might fall back in on itself in a "Big Crunch". At present astronomers cannot decide which scenario is correct. They say that the Universe is "open" if it will go on expanding forever and "closed" if the expansion will one day be halted. Between these two extremes there lies a critical condition where the Universe continually expands but at an ever-decreasing rate. It forever approaches a maximum radius but never actually reaches it, nor does it ever begin to contract. Figure 22.4 shows what is meant in the form of a graph of the relative size of the Universe against time.

The "Big Crunch" scenario allows the possibility of the Universe being reborn after all the matter enters the "celestial melting pot" at the end of the contraction phase. Maybe the Universe disappears into its own black hole, only to explode forth once more. It might then expand and contract in ever-repeating cycles like some form of cosmic yo-yo.

Exactly what happens to the Universe is very dependent on the amount of matter it contains. Taking account of everything that is visible, the indications are that the Universe does not contain enough mass to halt the expansion. In other words the Universe is open. However, astronomers are finding more and more evidence for invisible matter spread through the Universe. This "missing mass" betrays itself in the motions of stars in galaxies, and in a number of other ways. Many now think that as much as 99% of the matter in the Universe may be in a form that we cannot directly see. Much work still has to be done, but it now seems very likely that the missing mass is great enough to close the Universe. If so then the great cosmic clock will rewind. You never know – you might just be sitting where you are reading this book again in several trillions of years' time!

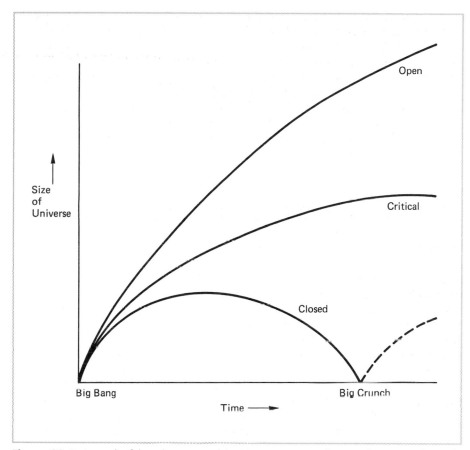

Figure 22.4 A graph of the relative size of the Universe vs. time, showing the "open", "critical" and "closed" scenarios.

Questions

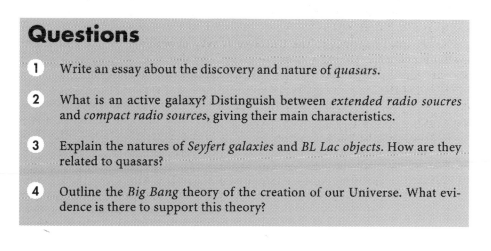

1. Write an essay about the discovery and nature of *quasars*.

2. What is an active galaxy? Distinguish between *extended radio soucres* and *compact radio sources*, giving their main characteristics.

3. Explain the natures of *Seyfert galaxies* and *BL Lac objects*. How are they related to quasars?

4. Outline the *Big Bang* theory of the creation of our Universe. What evidence is there to support this theory?

5 Discuss the possibilities for the eventual fate of our Universe.

6 Write down the standard form of the Doppler equation, as used to find the radial velocities of stars in our Galaxy and for nearby galaxies. Why cannot this equation be used for distant quasars? Give the equation you would use in determining the distances of quasars.

7 A quasar is observed to suddenly increase its brilliance over a period of a few months. An astronomer takes a spectrum of the quasar, identifies the spectral lines and measures their effective wavelengths in order to calculate the redshift value for the object. The astronomer finds that a line that is normally at a laboratory wavelength of 3×10^{-7} m has been displaced to a wavelength of 9×10^{-7} m.

(a) What is the redshift value, z?

(b) What fraction of its present size was the Universe when the brightening event occurred?

(c) Assuming that the Universe began 15 000 million years ago and has since been expanding at a uniform rate, how long ago did the brightening event actually occur?

(d) Assuming that Hubble's law is obeyed for the quasar and that the value of the Hubble constant is 65 km/s per megaparsec, how fast is the object receding from us? (1 parsec = 3.26 light years.)

(e) If Hubble's law really does hold true right out to the furthest limits of the Universe, what distance (in light years) would correspond to an object moving away from us at the speed of light? Comment on your answer.

Glossary of Terms

accretion disk   A disk of material formed around one astronomical body, often having been extracted from another.

albedo   A measure of the reflectivity of a planetary surface.

altitude   In astronomical usage, the angular distance separating a given celestial object from the horizon.

aperture synthesis   The process of combining the outputs of two or more small telescopes to mimic the effective output of a larger one.

aphelion   The position of a planet in orbit around the Sun when the planet–Sun distance is at its greatest.

apogee   The position of a body in orbit around the Earth when the Earth–body distance is at its greatest.

astronomical unit (AU)   The mean Earth–Sun distance; 1 AU = 150 million kilometres.

azimuth   The companion coordinate to altitude; the angular distance of a celestial body projected vertically downwards onto the horizon plane, measured eastwards from the north cardinal point.

Big Bang   The generally accepted scenario for the creation of the Universe.

binary star   Two stars mutually gravitationally bound into a co-orbiting system.

black body   A theoretical concept; a perfect emitter and absorber of radiation, obeying well-defined physical laws.

black hole   A region of space surrounding a collapsed star where the acceleration due to gravity exceeds the speed of light. In other words, the region inside the black hole is completely cut off from our Universe.

Bok globules   Small, relatively dense clouds of interstellar material, from which stars are thought to form.

celestial equator   The projection of the Earth's equator onto the celestial sphere.

celestial poles   The projection of the Earth's rotation poles onto the celestial sphere.

celestial sphere   An imaginary sphere having the Earth situated at its centre, on which the coordinate systems used by astronomers are projected.

Cepheid variables   Stars that are variable in brightness and which obey a specific law relating their brightness with the period of their variations.

Chandrasekhar limit   The maximum mass (1.4 solar masses) for a star to be able to exist as a white dwarf.

chromatic aberration   A defect of an optical system where the images of a point source in different wavelengths of light are brought to differing positions of focus.

chromosphere   The layer of matter situated above the photosphere of the Sun (or other stars), from which spicules and prominences originate.

cosmology   The study of the birth and evolution of the entire Universe.

declination   The angular distance separating a celestial body from the celestial equator, measured along a great circle that passes through the body and the celestial poles.

Doppler effect   The apparent change in wavelength of the radiation received from an emitting source caused by the relative motion between the receiver and the source.

eclipse   The passage of one celestial body behind another or into the shadow cast by it.

ecliptic   The observed plane of the Sun's yearly motion across the sky (caused by the Earth orbiting the Sun).

equinoxes   The two positions at which the ecliptic cuts the celestial equator.

Hertzsprung–Russell diagram   A plot of the luminosity (or magnitude) against spectral type (or temperature) for a collection of stars.

Hubble's law   An observed relationship between the recessional velocity of a galaxy and its distance.

interstellar medium   The material lacing the spaces between the stars.

Kepler's laws   Empirical laws describing how one celestial body orbits another. They have been given good mathematical foundations.

BL Lac objects   Galaxies with active nuclei having very bland spectra, dominated by continuum emission resulting from synchrotron processes.

librations   The apparent nodding and swaying motion of the Moon occurring over a lunar cycle.

light year   A unit of measurement equal to the distance a pulse of light would travel in one year.

luminosity   The brightness of a celestial body.

RR Lyrae stars   One type of variable star.

magnetopause   A boundary separating the region where the magnetic field of a planet is dominant over that of the rest of the Solar System.

magnitude   A logarithmic scale of the brightness of a celestial body. Apparent magnitude is a measure of the apparent brightness, while absolute magnitude refers to the body's real brightness.

Milky Way   The apparent band of radiance that crosses the sky owning to the concentration of stars along the plane of our Galaxy.

nadir   The point on the celestial sphere that lies vertically below the observer (and thus through the other side of the Earth).

nebula   A vast cloud of gas and dust in space formed by a concentration of the interstellar medium.

neutrino   An elementary particle liberated by nuclear reactions in the cores of stars (and during supernova outbursts). This particle has little or no mass and interacts very weakly with matter.

neutron star   A star that has become crushed under its own weight to such an extent that its electrons have fused with its protons to make neutrons. The star is mainly composed of this super-dense neutron material.

nova   An eruptive outburst in which a star dramatically increases its brightness. The result of mass transfer processes between the stars of a close binary system.

nucleosynthesis   The formation of new elements by nuclear reactions.

occultation   The passage of one astronomical body behind another, as a result of their respective motions.

opposition   When the Earth is exactly between the Sun and a planet of our Solar System, the planet is said to be at opposition.

parallax   The apparent shift in position of a celestial body against the backdrop of more distant bodies caused by the motion of the observer.

parsec   A unit of measurement, equal to 3.26 light years.

perigee   The position of a body in orbit around the Earth when the Earth–body distance is at a minimum.

perihelion   The position of a planet in orbit around the Sun when the Sun–planet distance is at a minimum.

photon   A "particle" of electromagnetic radiation.

photosphere   The visible surface of the Sun, or of another star.

precession   A slow shift in the orientation of the spin axis of a celestial body.

proper motion   The apparent shift in position of a star because of its real motion through space.

pulsar   A rapidly pulsing radio source. These have been identified with neutron stars.

quasar   Small and extremely powerful sources of electromagnetic radiation, thought to be the active nuclei of distant galaxies.

radial velocity   The component of a celestial body's velocity in the line of sight of the observer.

redshift   The lengthening in the wavelengths of the spectral lines from a celestial body caused by its recessional radial velocity (*see also* Doppler effect).

resolution   The discerning of fine details in an image.

right ascension   The companion of declination. It is the projection of a celestial body's position onto the celestial equator, measured from the vernal equinox in an easterly direction along the celestial equator. It can be measured as an angle or in units of sidereal time.

Roche limit   The distance from a planet at which any large body (such as a natural satellite) would break up owing to the shearing forces caused by the planet's gravitational field.

saros   The 18-year cycle of the Moon's motions such that after one Saros period the Earth, Moon, and Sun return to their original relative positions.

satellite   A small body orbiting a larger body.

Seyfert galaxy A type of galaxy having a nucleus that is optically very bright. It shows certain characteristics between ordinary galaxies and quasars.

sidereal period The orbital period of a planet moving around the Sun, or of a satellite moving around a planet.

sidereal time A system of time measurement based upon the apparent motions of the stars (caused by the rotation of the Earth).

solar apex The position on the celestial sphere which corresponds to the direction of travel of the Sun (taking its Solar System with it) with respect to nearby stars.

solar constant The total energy flux received per unit area per unit time at the radius of the Earth's orbit from the Sun.

solar wind A stream of electrified particles ejected from the Sun that spreads radially outwards through the Solar System.

spectral type The classification of a star based upon the characteristics of its spectrum.

supernova A colossal stellar explosion, resulting either from the collapse of the core of an old massive star, or from runaway nuclear reactions caused by the transfer of matter from a star to a white dwarf that are bound in a close binary system.

synchrotron radiation The emission of radiation resulting from electrons spiralling along magnetic field lines.

synodic period The period of one celestial body's motions around another with respect to the position of the Earth.

troposphere The lowest level in the Earth's atmosphere – the region of weather.

Van Allen belts Two concentric toroidal zones of solar wind particles trapped by the Earth's magnetosphere.

zodiac The band of sky through which the Sun, Moon, and all the planets (except Pluto) appear to travel in their orbits around the Sun. It is 16° wide, with the ecliptic passing through its centre.

Answers to the Numerical Questions

Chapter 1

8 Altitude of Polaris = 50°; (a)50°; (b) 25°. Declination of a star that will just not rise = –40°. Minimum declination of circumpolar star = +40°.

Chapter 2

9 (a) 16^h 37^m 27^s; (b) 09^h 09^m 27^s; (c) 62° 13′; (d) 49°.1 west of the meridian at 0^h UT and 79°.1 west of the meridian at 2^h UT; (e) 2^h 41^m UT on the following day (allowing 3 minutes for the difference in the sidereal and the mean time elapsed).

10 12^h 40^m.

12 (a) 22^h 23^m UT; (b) 32°.9; (c) due south.

13 (a) A solar day is longer; (b) 23^h 56^m; (c)(i) 00^h 00^m, (ii) 21^h 24^m UT.

Chapter 3

3 Focal ratio of 40 mm refractor = f/50.
Focal ratio of 160 mm refractor = f/12.5
Magnification of each telescope = ×200.

6 (a) ×112.5; (b) ×150; (c) ×80; (d) ×50; (e) ×400; (f) ×80. The combination used in (e) is unsuitable.

Chapter 5

1 (a) 780 million km; (b) 5.2 AU.

7 2.83 days.

Chapter 15

6 (a) X recedes with a velocity of 3.7×10^6 m/s.
(b) Y approaches with a velocity of 1.9×10^6 m/s
(c) Z recedes with a velocity of 6.3×10^6 m/s.

Chapter 16

1 1 : 19; star Y is the brighter.

2 (a) 14.7; (b) 370 mm.

4 (a) 2.33; (b) 29 parsecs; (c) 0.034 arc seconds; (d) Regulus is 209 times more luminous than the Sun.

Chapter 22

7 (a) 3; (b) $\frac{1}{4}$; (c) 11 250 million years ago; (d) 2.24×10^5 km/s; (e) 15 000 million light years.

Index